73 CRIME & DETECTION
74 RUSSIA
75 LIGHT
76 ENERGY
77 ELECTRICITY
78 FORCE & MOTION
79 CHEMISTRY
80 MATTER
81 TIME & SPACE
82 ASTRONOMY
83 EARTH
84 LIFE
85 EVOLUTION
86 ECOLOGY
87 HUMAN BODY
88 MEDICINE
89 TECHNOLOGY
90 ELECTRONICS
91 RENAISSANCE
92 IMPRESSIONISM
93 GOYA
94 MANET
95 MONET
96 VAN GOGH
97 WATERCOLOR
98 PERSPECTIVE
99 DANCE
100 FUTURE
101 MYTHOLOGY
102 LEONARDO & HIS TIMES
103 OLYMPICS
104 MEDIA & COMMUNICATION
105 TITANIC
106 FOOTBALL
107 HURRICANE & TORNADO
108 SOCCER
109 PRESIDENTS
110 BASEBALL
111 EPIDEMIC
112 WORLD WAR II
113 SUPER BOWL
114 CIVIL WAR
115 RESCUE
116 EVEREST
P9-CEZ-713

DORLING KINDERSLEY EYEWITNESS BOOKS

EVEREST

Buddhist wheel

Inca bag

Ski resort poster

Andean pot

Mountain goat

Snowboarder

EVEREST

Written by
REBECCA STEPHENS

A Dorling Kindersley Book

Tibetan Buddhist statue

Mountain eagle

Dorling Kindersley

LONDON, NEW YORK, SYDNEY, DELHI, PARIS, MUNICH, and JOHANNESBURG

Project editor Marek Walisiewicz
Art editor Rebecca Johns
Senior editor Monica Byles
Senior art editor Jane Tetzlaff
Category publisher Jayne Parsons
Senior managing art editor Julia Harris
Senior production controller Kate Oliver
Picture research Amanda Russell, Andrea Sadler
DTP designers Almudena Díaz, Matthew Ibbotson
Jacket designer Dean Price
US editors Margaret Parrish, Gary Werner

This Eyewitness ® Guide has been conceived by Dorling Kindersley Limited and Editions Gallimard

First American Edition, 2001
00 01 02 03 04 05 10 9 8 7 6 5 4 3 2 1

Published in the United States by
Dorling Kindersley Publishing, Inc.
95 Madison Avenue
New York, New York 10016

Ice-axes

Moche water pot from the Andes

Library of Congress Cataloging-in-Publication Data

Stephens, Rebecca.
Everest / by Rebecca Stephens.
p. cm. -- (Dorling Kindersley eyewitness books)
ISBN 0-7894-7395-X -- ISBN 0-7894-7398-4 (lib. bdg.)
1. Everest, Mount (China and Nepal)--Juvenile literature. [1. Everest, Mount (China and Nepal).] I. Title. II. Series.

D8495.8.E9 S73 2001
954.96--dc21

00-058939

Coin showing the head of Hannibal the Great

Color reproduction by Colourscan, Singapore
Printed in China by Toppan Printing Co. (Shenzhen) Ltd.

Mountain lion

Contents

Fan commemorating the ascent of Mont Blanc in the European Alps by Mr. Albert Smith in 1851

Mountains of the world

THE MOST SPECTACULAR of all landforms, mountains are found on every continent in the world. Together, they account for about one twentieth of the Earth's land surface. Some mountains stand as single peaks above the surrounding lowlands, but many more are part of long ranges that run for hundreds of miles, and series of ranges are sometimes grouped together into a cordillera. Individual mountains vary dramatically in size: those in Snowdonia, Wales, which struggle to reach 3,300 ft (1,000 m), would be mere foothills in the Himalayas and Karakoram, where 14 mountains top 26,000 ft (8,000 m). But whatever their size, mountains impose themselves on the landscape. They make their own climate, they are home to plants and animals not found in the lowlands, and they all radically influence the lives of people who live in their shadow.

ABOVE THE ALPS
A view from the sky shows part of a typical mountain range – the European Alps. The range takes its name not from the mountains themselves, but from the lower areas used for pasture in the summer. Passes through the Alps have been used for trade and warfare since prehistoric times.

A survey team uses a theodolite mounted on a sturdy tripod

MEASURING UP
The height of a mountain was once determined using a theodolite. This is essentially a telescope equipped with a level that allows the angle from one point (of known height) to another (the mountain summit) to be measured. The height of the mountain can then be calculated by trigonometry. Today, satellite navigation technology gives far more accurate measurements.

THIN AIR
In 1648, the French scientist Blaise Pascal carried a barometer up a mountain to show that atmospheric pressure drops with altitude. Pressure helps determine the density of the air that we breathe. At a height of 18,000 ft (5,500 m) pressure is half that at sea level. Climbers must adapt to the thin air during ascent.

There are 11 telescopes on or near the summit of Mauna Kea

THE HIGHEST MOUNTAIN ON EARTH
Mauna Kea, a dormant volcano in the Hawaiian islands, rises a modest 13,795 ft (4,205 m) above sea level, but measured from the ocean bed it is 31,988 ft (9,750 m) high – taller than Everest by 2,959 ft (902 m). The clean, unpolluted air at the summit makes it the ideal place for astronomical observations.

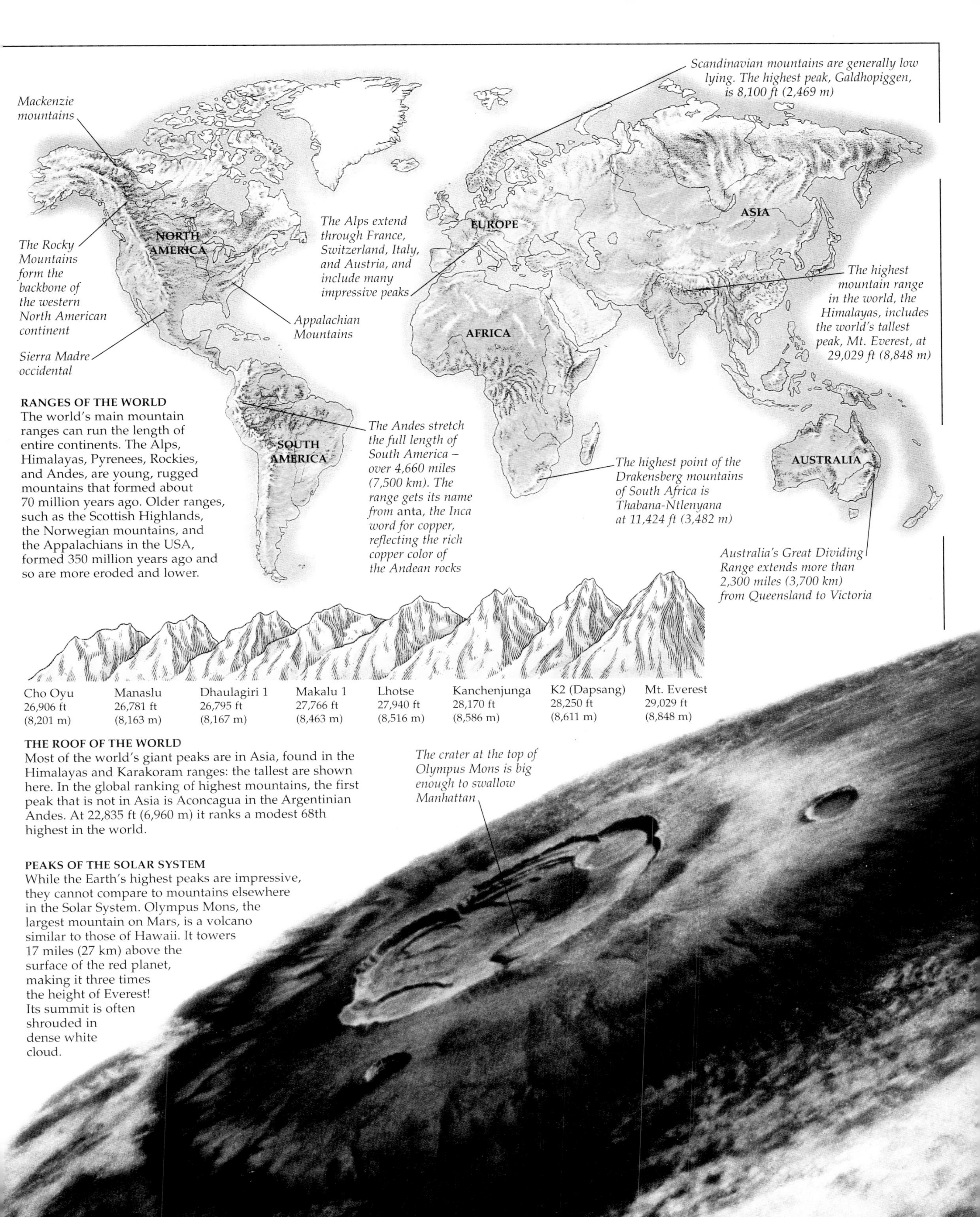

Scandinavian mountains are generally low lying. The highest peak, Galdhopiggen, is 8,100 ft (2,469 m)

The Rocky Mountains form the backbone of the western North American continent

Sierra Madre occidental

Appalachian Mountains

The Alps extend through France, Switzerland, Italy, and Austria, and include many impressive peaks

The highest mountain range in the world, the Himalayas, includes the world's tallest peak, Mt. Everest, at 29,029 ft (8,848 m)

RANGES OF THE WORLD
The world's main mountain ranges can run the length of entire continents. The Alps, Himalayas, Pyrenees, Rockies, and Andes, are young, rugged mountains that formed about 70 million years ago. Older ranges, such as the Scottish Highlands, the Norwegian mountains, and the Appalachians in the USA, formed 350 million years ago and so are more eroded and lower.

The Andes stretch the full length of South America – over 4,660 miles (7,500 km). The range gets its name from anta*, the Inca word for copper, reflecting the rich copper color of the Andean rocks*

The highest point of the Drakensberg mountains of South Africa is Thabana-Ntlenyana at 11,424 ft (3,482 m)

Australia's Great Dividing Range extends more than 2,300 miles (3,700 km) from Queensland to Victoria

THE ROOF OF THE WORLD
Most of the world's giant peaks are in Asia, found in the Himalayas and Karakoram ranges: the tallest are shown here. In the global ranking of highest mountains, the first peak that is not in Asia is Aconcagua in the Argentinian Andes. At 22,835 ft (6,960 m) it ranks a modest 68th highest in the world.

The crater at the top of Olympus Mons is big enough to swallow Manhattan

PEAKS OF THE SOLAR SYSTEM
While the Earth's highest peaks are impressive, they cannot compare to mountains elsewhere in the Solar System. Olympus Mons, the largest mountain on Mars, is a volcano similar to those of Hawaii. It towers 17 miles (27 km) above the surface of the red planet, making it three times the height of Everest! Its summit is often shrouded in dense white cloud.

Mountains in the making

EVERY YEAR, THE WORLD'S HIGHEST MOUNTAIN RANGE, the Himalayas, grows by about 0.5 in (1 cm). The huge forces in the Earth's crust needed to build mountains come from the slow movement of the giant plates on which our planet's continents and oceans sit. The Himalayas began to form about 50 million years ago when India, then an island, was carried northward on its continental plate and crashed into Asia. Where the two collided, the rock buckled and was pushed upward, and the Himalayas were born. Because India continues to push into Asia, at a rate of 2 in (5 cm) a year, the mountains are still growing upward. Other mountains form from different types of plate movements. For example, a collision may cause lava (molten rock) to be ejected from deeper levels within the Earth, making a volcanic mountain.

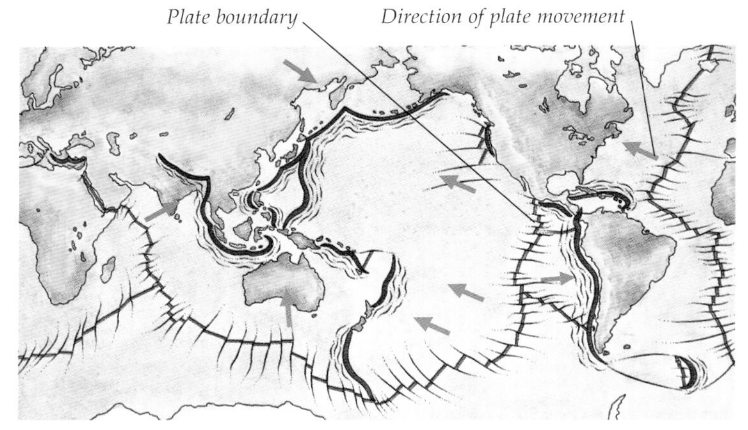

PLATES, COLLISIONS, AND MOUNTAIN PEAKS
The Earth's shell, or lithosphere, is split into nine large plates and about 12 smaller ones. Continents are embedded in continental plates, while oceanic plates make up the sea floor. The plates move at about 0.5 in (1 cm) per year – or 6 miles (10 km) every million years. Some move toward one another, or apart, and others slide past each other sideways. Younger mountains tend to be found at the edges of continents because it is here that the oceanic plates crash into continental plates.

WEATHERED GIANTS
As soon as a mountain forms, the forces of erosion begin to grind and weather its surface, reducing its height over the years. As a result, the highest mountains in the world also tend to be the youngest. The Appalachian Mountains, which run through North America from Newfoundland to Alabama, are an old range, formed around 250 million years ago. Once they were as tall and impressive as the modern Andes, but today they have been eroded into gentle hills.

MOUNTAIN BUILDING

Almost all mountains form at the boundaries of plates that make up the Earth's surface. Where plates collide, thick lava, dust, and rock fragments can be forced out, making a steep cone-shaped volcanic mountain. Sometimes the rock buckles as it is compressed by colliding plates, making fold mountains. When there is compression or tension in the rock, an area of crust can be lifted up or drop between faults (lines of weakness) in the bedrock, forming block-fault mountains.

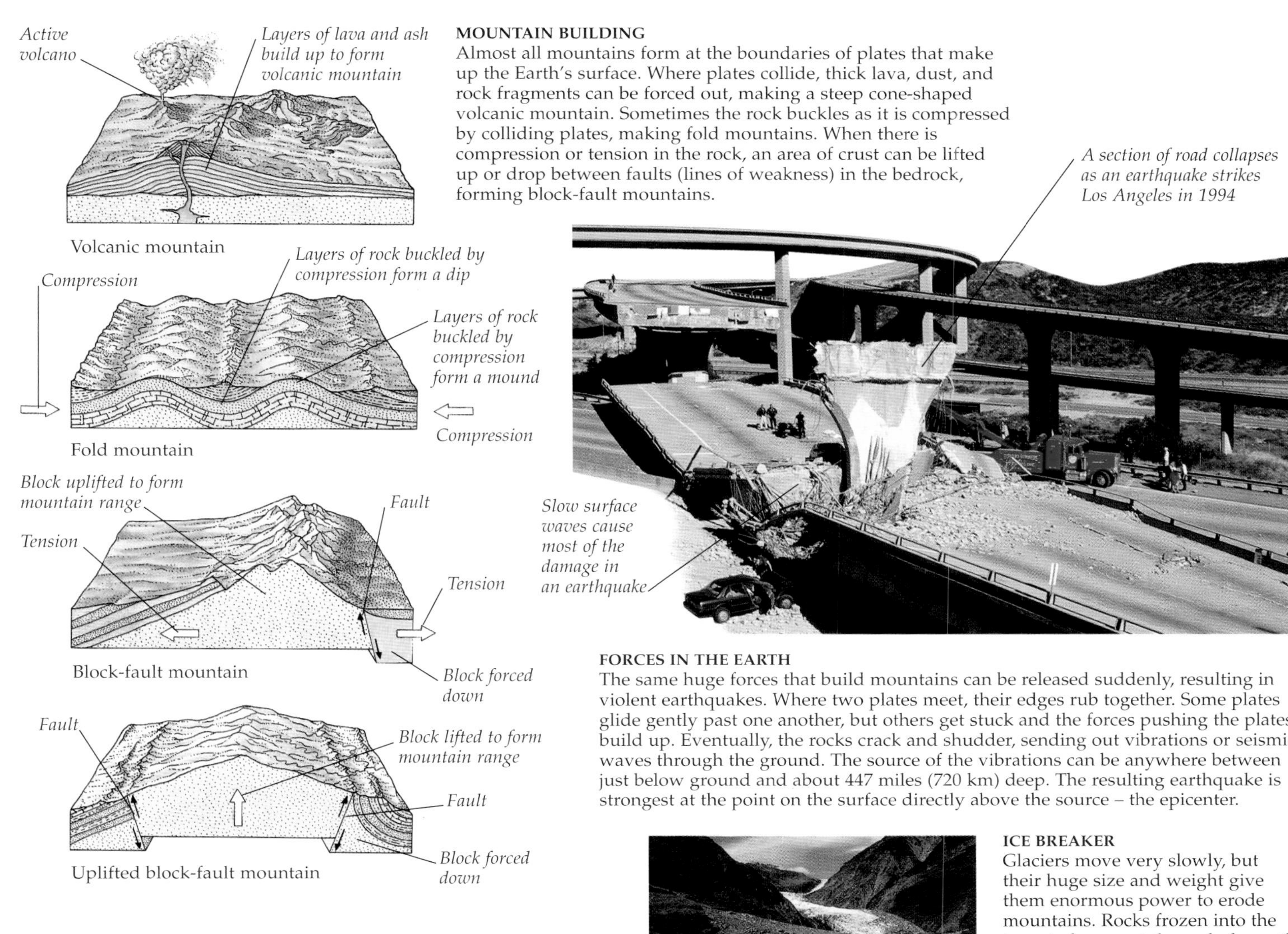

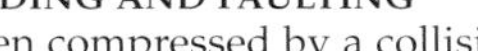
Uplifted block-fault mountain

FORCES IN THE EARTH

The same huge forces that build mountains can be released suddenly, resulting in violent earthquakes. Where two plates meet, their edges rub together. Some plates glide gently past one another, but others get stuck and the forces pushing the plates build up. Eventually, the rocks crack and shudder, sending out vibrations or seismic waves through the ground. The source of the vibrations can be anywhere between just below ground and about 447 miles (720 km) deep. The resulting earthquake is strongest at the point on the surface directly above the source – the epicenter.

ICE BREAKER

Glaciers move very slowly, but their huge size and weight give them enormous power to erode mountains. Rocks frozen into the ice grind away at the rocks beneath, and the glacier transports the rock fragments to its base, where they accumulate in a moraine. A glacier only a few hundred yards across can tear up and crush millions of tons of rock a year.

FOLDING AND FAULTING

When compressed by a collision of plates, the rocks of the Earth's crust begin to fold. The more they are squeezed, the more folded they become. Simple folds may become overturned (nappes), and faulting occurs, breaking the crust into sheets of rock, or thrust sheets, up to 12.4 miles (20 km) thick. This complex folding and faulting absorbs the force of plate collision over an area that extends hundreds of miles into the continent, and is recorded in the banded rock strata, here simulated using colored sand.

The Seven Summits

THE HIGHEST MOUNTAINS on each of the world's seven continents – Africa, Asia, Europe, North America, South America, Antarctica, and Australia – are known by mountaineers across the world simply as "The Seven Summits." Each peak is very different from the others, and climbing all seven has become a challenge for adventurers since the 1980s. The highest point in Australia is Kosciusko, a hill that provides no real challenge, so some argue that another much higher mountain, Carstensz Pyramid, on the island of New Guinea just north of Australia, should take its place in the seven.

The climate on Mt. McKinley's summit is colder than at the North Pole

MT. MCKINLEY
At a grand height of 20,320 ft (6,194 m) and just a few degrees below the Arctic Circle, McKinley is one of the coldest peaks in the world and the tallest in North America. Its massive, snowy summit rises high above the Alaskan plain. McKinley is extremely challenging because of its Arctic conditions, frequent avalanches, and deep crevasses: even today, only half of the expeditions to the mountain reach the summit.

Dick Bass unfurls a stars and stripes banner on climbing the seventh summit

THE SEVEN SUMMITS SCALED
On April 30, 1985, the Texas entrepreneur and property developer Dick Bass stood on the summit of Everest. He had just completed his four-year globe-trotting quest to become the first person to climb the highest peak on the world's seven continents. The following year, in August 1986, a Canadian photojournalist called Pat Morrow became the first person to climb an alternative version of the Seven Summits, which includes Carstensz Pyramid, New Guinea, in place of Kosciusko in Australia.

In 1992, 32 climbers stood on Everest's summit in a single day

MT. EVEREST
Everest straddles the border of Tibet and Nepal in the Himalayas. At 29,029 ft (8,848 m), it is the highest mountain in Asia and the world, and attracts great prestige to anyone who reaches its summit. The Tibetan name for Everest is Chomolungma (Goddess Mother of the World); its Nepalese name is Sagarmatha (Goddess of the Sky); and in the West, it was known for many years as Peak XV before it was recognized as the world's highest summit.

KILIMANJARO
The highest peak in Africa, Kilimanjaro, is an extinct volcano that stands just south of Tanzania's border with Kenya. It is only a stone's throw from the equator, and its huge, sprawling base is surrounded by lush alpine forests. Because of its height of 19,340 ft (5,895 m), Kilimanjaro has a glacier and a cap of snow, which is often mistaken for clouds floating above the surrounding arid plain. The mountain has three summits called Mawenzi, Kibo, and Shira.

MT. ELBRUS
Many people think that Mont Blanc on the border between France and Italy is the highest mountain in Europe. In fact Mt. Elbrus, at 18,510 ft (5,642 m), is higher by almost 3,000 ft (1,000 m) than any peak in the Alps. Elbrus – an extinct twin-headed volcano – lies in the Caucasus on the very southern edge of Russia, between the Black and Caspian seas.

ACONCAGUA
At 22,834 ft (6,960 m) high, Aconcagua is the highest mountain in South America and the second highest of the Seven Summits. The wedge-shaped giant lies in the Andes, just inside Argentina's border with Chile. The Incas called it "White Sentinel" because it rises so dramatically from the stony valleys below.

CARSTENSZ PYRAMID
This mountain vies with Kosciusko as the highest peak of Australasia, but most mountaineers agree that it is more interesting and challenging to climb. Surrounded by rain forest in Irian Jaya, New Guinea, it is a rugged limestone crest of 16,024 ft (4,884 m) and the only one of the Seven Summits that requires any skill in rock climbing.

MT. VINSON
Until recent times, Vinson, the highest mountain in the icy wilderness of Antarctica, was all but inaccessible. In fact, Vinson was the last continental summit to be discovered and conquered by mountaineers. Now it is possible to fly to Antarctica, and take a small airplane to the foot of the mountain. At 16,023 ft (4,897 m), it is of no great height, but it is thought to be one of the most beautiful of all peaks. The extreme cold, dryness, and desolation of the environment make climbing Vinson a challenging undertaking.

Mountain features

FROM A DISTANCE, mountains can look almost featureless – shadowy forms stretching out along the horizon. But get among them and a whole world of glaciers and gullies, ridges and buttresses, cirques and cols opens up. Mountains are complex landforms, and are changing all the time. Just as soon as a mass of land is pushed up above sea level it is attacked by wind and rain, eroded by streams and rivers, and attacked by ice and frost. In deserts, mountain rocks are broken apart by huge extremes of temperature, producing structures that challenge scientists and mountaineers alike.

SLOW RIVERS OF ICE
Snow falls and settles high up in cold mountains and may not melt away in warmer months. Over the years, it piles up in hollows, or cirques, and is compressed into ice. Eventually, the ice becomes so heavy that it begins to flow slowly down the mountain. A river of ice, or glacier, is formed. As the glacier moves, it picks up rubble, which scrapes against the mountain, cutting a U-shaped valley into the rock.

Cirque (corrie)

Ridge

Tributary glacier

Medial moraine (rocks and debris)

Firn (compressed snow)

Moving ice

U-shaped valley carved by glacier

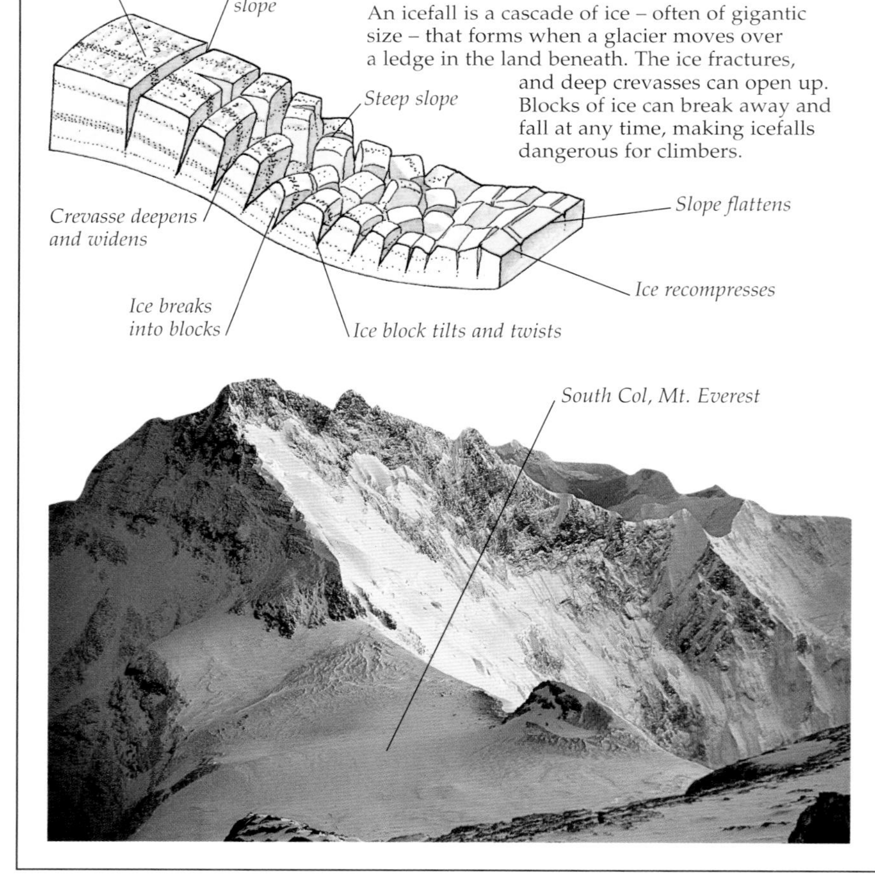

ICEFALLS
An icefall is a cascade of ice – often of gigantic size – that forms when a glacier moves over a ledge in the land beneath. The ice fractures, and deep crevasses can open up. Blocks of ice can break away and fall at any time, making icefalls dangerous for climbers.

DANGER – CREVASSE!
A crevasse is a split in a glacier, which may be very deep. Here mountaineers climb a crevasse on ladders. Sometimes, fresh snow collects around the top of a crevasse hiding it from view. Climbers are acutely aware of the danger of these hidden traps, and rope together so that if one of a group is unlucky enough to fall in, the others can hold the rope and save their companion from serious injury.

EVEREST'S SOUTH COL
A col is a saddle-shaped depression in a mountain chain, and none is as famous as the South Col on Everest, here viewed from above, from the slopes beneath the summit. The South Col is a place that evokes mixed feelings among climbers, because it is where many expeditions pitch their highest camp. It is a hostile, windswept, and lonely place to be, and yet it is also the launch pad to the summit.

AT THE SUMMIT
A mountain summit may be sharp and rugged, like that of Everest, dome-shaped, like that of Mont Blanc in the European Alps, or even a broad plateau.

GULLIES
A gully is a wide split in a mountainside typically formed by the erosive action of material carried in a stream or river. Gullies can be dry, as here, or coated with snow and ice. Either way they provide an interesting challenge for a climber.

VERTICAL WALLS
Steep faces of rock or ice were once impossibly difficult obstacles for climbers. With modern equipment and advanced techniques, climbers can tackle vertical walls and even overhangs.

Viscous flowing ice

Crevasse

Ribbon lake

Push moraine

Sediment deposited by meltwater

Terminal moraine

Outwash plain

Meltwater

Stream

Wind and weather

MOUNTAINS CREATE THEIR OWN WEATHER, which is usually colder, windier, and more unpredictable than in the valleys and plains that surround them. Temperatures plummet with altitude, with –94°F (–70°C) often recorded on the summit of Everest, and winds can reach 192 mph (320 km/h) because they gain speed as they rush over mountains. Most heat from sunlight is reflected back up into the atmosphere by the white ice and snow on mountain peaks, and the northern slopes of many mountains are in perpetual shadow, making them especially cold. The very presence of mountains is a barrier to the normal movement of air. As moist air rises to pass over a mountain range, it cools, and the moisture it holds condenses and falls as rain or snow. Typically, this makes the windward side of a mountain very wet and misty, and the sheltered side dry. In this way mountains not only create their own weather but also affect conditions over vast tracts of land that fall in their shadow.

STORMS OVER THE SUMMITS
Mountains are known for their dramatic thunderstorms, which occur when moist air rises quickly and becomes unstable. Ice crystals and raindrops move around violently in the clouds and, because of the motion, an electric charge builds up, which is discharged in the form of lightning.

SNOW-EATING WINDS
When moist air from over the Pacific Ocean moves inland over the United States, it meets the Rocky Mountains. Here it is forced upward into areas of lower atmospheric pressure, where it expands and cools. The moisture it carries condenses and falls as rain or snow. The air then flows down the eastern slopes of the Rockies, warming up rapidly. This warm wind is called the Chinook or "snow eater" because it can quickly melt snow on the ground.

WIND WARNING
Weather stations are often found on mountain peaks, because scientists need to study the most extreme weather conditions on our planet. Wind speed is measured using a spinning cup anemometer, which was invented in 1846. The strongest wind on the surface of the Earth – a gust of 231 mph (372 km/h) – was recorded on the summit of Mount Washington in New Hampshire in April 1934.

HOSTILE HEIGHTS
Both temperature and atmospheric pressure fall with increasing altitude. As mountaineers climb higher, they are less able to deal with the cold because they are starved of the oxygen they need. This means that mountaineers are very susceptible to cold-related problems such as frostbite and hypothermia.

JET STREAM
The jet stream is a current of fast-moving air that travels from east to west in the upper levels of the atmosphere, somewhere between 25,000 and 45,000 ft (7.5 and 14 km) above the Earth's surface. The summit of Everest, at 29,029 ft (8,848 m), is sometimes clipped by the jet stream. When this happens, mountaineers on Everest hear a deafening roar, like a freight train passing. Here the effect of the jet stream can be seen in the form of a plume of cloud streaming off the jagged ridge of Nuptse, a mountain that is a part of the Everest system.

SWIRLING BLIZZARDS
Wind can be a mountaineer's worst enemy. Wind speed tends to increase with altitude, so not only does the climber risk being physically blown off a high ridge by strong gusts, but he or she also suffers from the cooling effect of wind (windchill). Wind can whip up falling snow and ice grains on the ground into a swirling blizzard that reduces visibility to near zero.

A cornice is a bank of snow shaped like a wave. It projects over the edge of a ridge, and can collapse at any time

SNOW BUSINESS
Fresh, cold snow that is too soft to support a climber's weight is called powder. It is light and fluffy because there is a lot of air between the individual snow crystals. It is sought after by skiers but makes for deep drifts that are difficult to walk through. Firn snow, which has been compacted by partial melting and refreezing, is easier to walk over.

Life at the top

THE HIGHEST MOUNTAIN PEAKS are hostile environments. At altitudes of more than 16,400 ft (5,000 m), few plants and animals can survive the extreme conditions of freezing cold, scouring icy winds, thin air, and intense sunlight. But descending from the icy summits, the temperature rises by about 1°F (0.5°C) for every 330 ft (100 m) drop, and the mountain comes to life. Because conditions change greatly with altitude, a mountain can support a far wider variety of plants and animals than a similar area of flat land. The different faces and features of a mountain, such as its ridges, glaciers, and gullies, also create microclimates that provide a mosaic of habitats for a huge diversity of life.

INSECTS IN THE ICE
Rock crawlers are high-altitude specialists. Related to crickets, these insects live on mountain snowfields or alongside glaciers, and die at temperatures of more than 50°F (10°C). They feed on any organic material available at these heights, and grow slowly, taking seven years to complete a generation.

The wingspan of an Andean condor can reach 10 ft (3 m)

HIGH FLYER
The Andean condor makes its home among the highest mountains in the world. It breeds only at altitudes of more than 10,000 ft (3,000 m) in the high Andes, well out of reach of potential predators. It is one of the largest and heaviest flying birds, weighing about 22 lb (10 kg), and can soar at heights of 26,000 ft (8,000 m), near the cruising altitude of a passenger airliner.

ZONES OF LIFE
From the tropics to the Arctic Circle, the different levels of altitude on a mountain create distinct horizontal zones of life on its slopes. In the mountains of North America, permanent snow and ice cover on the peaks gives way to tundra, home to hardy lichens. On lower slopes lie meadows carpeted with cold-tolerant flowering plants, grasses, sedges, and rushes, and beneath these a belt of dwarf shrubs and conifers.

A rocky ridge offers a different habitat from a sheltered gully

Conifers have narrow leaves with a waxy coating that helps to retain water

Melting ice and snow water the meadows in spring

THE TREE LINE
Trees cannot survive above a certain height on a mountain. The closer the mountain lies to the equator, the higher this limit, or tree line, because the climate is warmer. Individual trees may grow above the normal tree line if sheltered within a gully.

MOUNTAIN ACROBATS
Mountain goats live on craggy peaks, usually above the tree line, feeding on mosses, lichens, and scrub foliage. They thrive because they are agile and sure-footed, able to leap more than 10 ft (3 m) to reach a patch of vegetation. The goats also face fewer predators at high altitude.

CLINGING ON FOR DEAR LIFE
Exposed rocks and trees on high mountain slopes are typically cloaked in lichens. Although they look like plants, lichens are partnerships between algae and fungi. They draw the water and nutrients they need to grow and reproduce from moisture in the air, and so are very sensitive to airborne pollution. The clean air of mountain regions provides an ideal habitat.

Yak from the Himalayas

Mountain specialists

Flight feathers splay out like fingers at the tips of the wings

MOUNTAINS ARE ISLANDS of life surrounded by seas of lowlands. Plants and animals that inhabit the upper reaches of one mountain cannot easily cross the hostile lowlands to colonize another peak – they are stuck in their small, closed community. Isolation means that new species evolve relatively quickly; and as a result, mountains are home to many rare, unique, and specialized plants and animals. For example, snow leopards, which live at altitudes of up to 20,000 ft (6,000 m) in the Himalayas, are well adapted for hunting in the snow: their legs are angled in such a way that they can jump 50 ft (15 m) vertically. Yaks, also from the Himalayas, breathe more slowly than their lowland relatives, cattle, and have more red blood cells to help them absorb oxygen from the thin mountain air.

THE LAST GORILLAS
The spectacular volcanic mountains between Congo, Zaire, and Uganda are the last refuge of the mountain gorilla. At up to 6 ft (1.8 m) in height, and weighing more than 400 lb (180 kg), these animals appear ferocious but are actually rather shy and feed only on vegetation. They live in close-knit family groups led by an adult male called a silverback because of the gray hairs on his back. Despite its remote and inaccessible habitat, the mountain gorilla is threatened by development and human activity, and only 630 now remain in the wild.

Distinctive white V on chest

MOUNTAIN RAIDER
Human activity threatens the Himalayan black bear, which lives in forested mountains at heights of 4,000–12,000 ft (1,200–3,600 m). In winter, the bear descends into valleys taking livestock, such as sheep and goats. This brings it into conflict with people who hunt the unwelcome visitor.

SOARING PREDATORS

Eagles can glide high over mountain slopes by rising on thermals – warm updrafts of air. With their keen eyes focused on the land below they can spot the small movements of prey. They also feed on the carcasses of animals that have slipped and fallen on the steep slopes. Verreaux's eagle hunts in the rocky hills of Africa and the Middle East, taking mainly hyraxes and other small mammals.

Red algae grow and reproduce within the ice of a glacier

Strong breast muscles power the eagle's flight

LIFE IN A FREEZER

Plants can exploit the harshest of mountain habitats – glaciers. Red algae are able to grow within the ice sheet: their red pigment allows them to convert light into heat and also filters out damaging radiation from the sun.

The ptarmigan's toes are covered with stiff feathers above and below for insulation

Thyme

COLOR CHANGE

While many mountain birds migrate to warmer climates in the winter, the ptarmigan stays on its mountain home all year round. In winter, its plumage changes from a gray-brown color to pure white so that it is well camouflaged against the snow.

CUSHIONED FROM THE COLD

Mountain plants typically grow in low, dense cushions to protect them from freezing conditions at high altitude. Fine hairs on their leaves trap moisture and warmth, and help protect the living tissues from intense sunlight.

Lynx

THE AGILE LYNX

The Spanish lynx is particularly well adapted to nocturnal life on the mountain. Its dense, soft fur provides good insulation, and it is an agile climber and swimmer.

An adult puma may reach a length of 9 ft (3 m) from head to tip of tail, and weigh 220 lb (100 kg)

The purr of a puma is like that of a domestic cat, but much louder

MOUNTAIN LION

The puma, also called the cougar or mountain lion, lives on mountains and in wilderness areas from British Columbia in Canada to Patagonia. It hides in rocky places, well camouflaged by its pale brown coat. It takes prey as large as deer, and has even been known to attack humans.

GURKHA HAT
Gurkha soldiers from Nepal have served in the armies of Britain and India for many years. They are admired for their strength and courage. Many return home after service abroad to become teachers and community leaders.

People of the Himalayas

THE HIMALAYAS GET THEIR NAME FROM the ancient Sanskrit words *hima* (meaning "snow") and *alaya* (meaning "abode"). The mountains stretch over 1,550 miles (2,500 km) from east to west, and lie for the most part in India, Nepal, and Bhutan, although they also extend into Pakistan and China. The inhabitants of this vast range are very diverse. Some are descended from European groups from the west, some from Indian people of the south, and others from Asiatic tribes of the north and east. They are united by the challenge of making a living in the highest mountain range in the world. Since the 1940s, this challenge has been met very successfully, with the result that growing populations are placing great strain on the fragile mountain environment.

Most Bhutanese people have a simple but healthy diet including potatoes, rice, and chiles

BHUTANESE FAMILY
The Kingdom of Bhutan lies on the eastern edge of the Himalayas. The majority of its inhabitants are descended from Tibetan people who settled in the area in the 9th and 10th centuries. Most are devout Buddhists, and their language, Dzongkha, is very similar to that spoken in Tibet. A large minority of people in Bhutan are recent immigrants from Nepal.

Bhutanese crafts include fine fabrics and jewelry

SHERPA CAPITAL
Namche Bazaar is the main town in Solu Khumbu, the region around Mt. Everest where the majority of Sherpa people live. Built in a natural amphitheater at 11,300 ft (3,440 m) and surrounded by high hills, Namche Bazaar has no cars – the only way in is by foot. Despite this, it has stores, restaurants, a bakery, hotels with hot showers, and even a bank. The colorful market takes place every Saturday.

DECORATED TRAVEL
Ornate trucks and buses are a common sight throughout India and the Himalayas. Roads run into the Himalayas from both north and south, although they need constant maintenance to repair the damage caused every year by torrential monsoon rains. The steep terrain makes rail transportation almost impossible, although an electrically-powered aerial tramway is used to ferry goods into Kathmandu, the capital of Nepal.

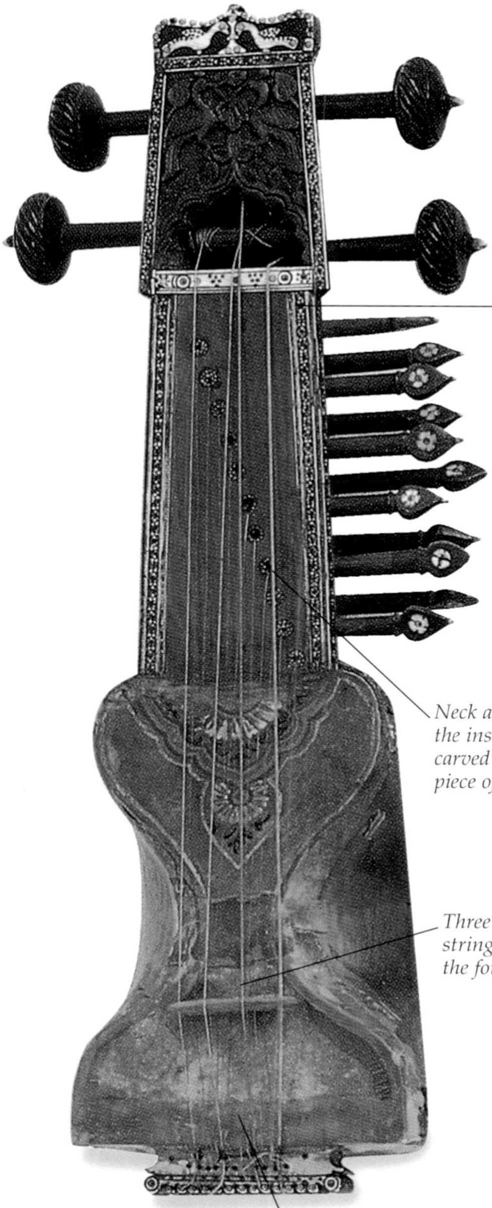

SARANGI

The classical music of Nepal and northern India is marked by the shimmering sounds of exquisite instruments like the *sarangi*. Stringed instruments are used with drums and wind instruments in both religious and secular celebrations.

Neck has no frets. Eleven strings pass through the fingerboard, vibrating in sympathy with the four main strings to produce a more complex sound

Neck and body of the instrument are carved from a single piece of wood

Three of the four main strings are made of gut – the fourth is metal

The sarangi *is held upright and played with a bow*

CHARIOTS OF THE GODS

Scores of colorful festivals take place every year in Nepal's Kathmandu Valley. This festival, called Bisket Jatra, takes place in April in the old kingdom of Bhaktapur. Images of demons and gods are placed on giant chariots and dragged around the town. Wherever they stop, people make offerings of coins, flowers, and even blood to the deities.

Continued on next page

Continued from previous page

Customs and religion

Squeezed between the giant countries of China and India, the people of the Himalayas have developed cultures that fuse many, diverse influences. Nepal, for example, is officially a Hindu country, but the religion practiced by most inhabitants incorporates Buddhist beliefs and Tantric gods.

Many sadhus shave their heads or grow their hair long as an act of devotion to the god Shiva

Markings on the sadhu's forehead identify the sect to which he belongs

Garlands of beads represent the elements of creation

ROADS TO ENLIGHTENMENT
For Hindus, spiritual enlightenment is the highest goal in life. Sadhus, or holy men, renounce the world in order to focus entirely on the Higher Reality beyond. They cut all family ties, have no possessions or home, wear little or no clothing, and eat simple food. Usually they live by themselves and spend their days in devotion to their chosen deity. Sadhus perform magic rituals, practice yoga, or play music to make contact with the gods.

SHIVA, CREATOR AND DESTROYER
Hindus have many gods, but consider them all to be aspects of one deity. Each god has a special set of features – they may carry a weapon, or have additional limbs or animal characteristics that relate to their special powers. Hindus usually worship one favorite god. One of the most important is Shiva, god of destruction and creation. Here a statue of Shiva is covered with pigment and ashes, symbolic of death and regeneration.

HOLI MARKS
During the Hindu festival of Holi in the spring, people remember the god Krishna and his beloved companion Radha. Worshippers cover themselves with *gulal* – brightly colored powder paints made from vegetable pigments – and light fires to rid the air of evil spirits.

BUDDHIST BUILDING
A stupa is a hemispherical structure that sums up Buddhist belief. Shown here is the Bodhnath stupa in Kathmandu, Nepal – one of the largest in the world. The base of the stupa takes the shape of a mandala (the Sanskrit word for "circle"), which symbolizes earth. On this sits the dome, which represents water. Then comes the spire, symbolizing fire, and the pinnacle, which stands for ether. The Buddha's ever-watchful eyes stare out in four directions from the square base of the spire.

Vairochana perfects knowledge

Amitabha, "Infinite Compassion," perfects speech

Vajrasattva, "The Unchanging," perfects wisdom

Amoghasiddhi, "Almighty Conqueror," perfects action

Ratnasambhava, "The Beautifier," perfects goodness and beauty

FIVE BUDDHAS
This headdress belonging to a Tibetan lama (priest) shows the Buddhas of Meditation, which, according to belief, live in the heavenly worlds. Each personifies an aspect of "Divine Being" that meets a dead person's spirit. The spirit's reaction shows how enlightened the person is and determines how he or she will be reborn.

Tibetan shaman's headdress

ANCIENT SORCERY
A minority of people in the Himalayas believe in the powers of the shaman – a priest or medicine man. The shaman is believed to accomplish feats of healing and spiritual guidance by putting himself into a trance. During the trance he is thought to leave his body and enter the spirit world, from where he can return with divine wisdom.

Drum and bell are devotional instruments

THE LIFE OF A MONK
Buddhist monks in the Himalayas live according to strict moral guidelines. Traditionally, they live as wandering beggars, settling in one place only for the three months of the rainy season.

Traditional saffron robes of Buddhist monk

European mountain people

THE CONTINENT OF EUROPE is criss-crossed by mountain ranges, some worn down over centuries by glacial action, some continuing to form, with live volcanoes. The small states of Andorra, Liechtenstein, and Monaco lie entirely in the mountains. A long belt of young mountain ranges stretches across southern Europe containing Europe's highest and most rugged mountains, from the Sierra Nevada in Spain across to the Balkan Mountains in the east. Although modern tourist resorts bring wealth to some nations, many mountain communities follow a centuries-old way of life.

STORING GRAIN
Traditional granaries made of stone or wood, called *espigueiros*, are still used in some mountainous areas of Portugal. Their design has hardly changed since the 18th century. Raised on columns, they allow grain to be stored at the right humidity and away from rodents.

MOUNT ETNA
At around 10,500 ft (3,200 m), Mount Etna in Sicily is the highest active volcano in Europe. Despite the risk of eruptions, farming communities inhabit the fertile lower slopes cultivating vines, olives, and citrus fruits.

Tent used by plant gatherers in the Alps

PLANT COLLECTORS
During the 18th and 19th centuries, many amateur naturalists visited the mountainous regions of Europe to gather plant specimens. Such collectors often created alpine gardens for their homes using samples carefully brought back from their travels.

Hairy leaves

Mountain kidney vetch

Spring gentian

Flowers appear from April until June

NATURE-LOVING VISITORS
The beauty and abundance of wild alpine flowers attract many visitors to the mountains of Europe. During the short summer season, nature enthusiasts travel hundreds of miles to enjoy the rich variety of plants that grow in grassy meadows on the lower slopes and study specialized plants that thrive at higher altitudes.

Swiss farmer in traditional costume

DAIRY FARMING
The production of milk, butter, cheese, and yogurt is an important source of income for some mountain communities. Farmers in the Swiss Alps rely on cows for dairy products, while in mountainous parts of southern Europe, goats and sheep provide alternative supplies of milk.

FJORDS OF SCANDINAVIA
Some of the most breathtaking mountain ranges in Europe are found in Scandinavia. In the Norwegian fjords, dramatic cliffs beside the water's edge rise up almost vertically for hundreds of feet. More than half of Norway is mountainous, which made long-distance travel difficult until relatively recently. During the winter months, when snow falls are heavy, roads become impassable and railroad tracks may be closed. Ferries play a key role in public transport, and shipping is a major industry.

A Norwegian fjord surrounded by mountain peaks

Lords of the Andes

Wooden Inca cup portraying Francisco Pizarro, conqueror of Peru

THE INCA EMPIRE WAS ONE OF THE BEST organized mountain civilizations in history. Extending nearly 2,500 miles (4,000 km) from the northern border of modern Ecuador to central Chile, it once had control over more than 12 million people living from the Pacific coast to the highlands of the Andes. Until about AD 1438, the Incas were just one of several groups of people living in the southern Andes. But then, under their ruler Pachacuti, they set out to conquer the land and the people around them, expanding their territory to the north, south, and west. At the peak of the empire, the Incas lived in a highly developed society; but its success was short lived. In 1532, a small Spanish army, led by a general named Francisco Pizarro, invaded the Incas, killed their ruler, and took over the empire, pocketing a large amount of gold for Spain in the process.

GOLD GIFTS
The Incas were famous for crafting metals, such as copper, silver, platinum, and gold. Female figures like this one have been found among their offerings to the gods.

PEAK PRACTICES
From the 1st to the 8th centuries AD, the dominant civilization of the Andean region was the Moche. These people built large cities and developed systems of irrigation channels that carried water from the high Andes to their crops of corn and beans. They were skilled artisans, producing sophisticated and beautiful jewelry and pottery.

Moche water jars were often decorated with depictions of people, animals, or demons

The people of the Andes were typically small, with straight black hair, high cheek bones, and brown skin

Machu Picchu is the biggest single tourist attraction of modern Peru

MACHU PICCHU
Strategically positioned at the edge of the Inca empire, the remote city of Machu Picchu is a stunning example of Inca architecture – a natural fortress protected by steep slopes, and approachable from only one point. Of its 143 granite buildings, about half were houses and the rest ceremonial buildings, such as temples.

HIGH HERDSMEN
Llamas have been valued for centuries as pack animals. They are also a source of food, wool, and hides, and their dried dung can be used as fuel. Llamas are hardy creatures that can go for a long time on little water, making them ideal stock to keep in the rugged, dry Andes.

Continued on next page

Continued from previous page

Inca culture

Although the Incas left no written records, much can be learned about their way of life by examining their architecture, pottery, and clothing. These people believed that, after death, they would carry on living in another world, so they took many belongings with them to their graves. These items, especially the colorful textiles made from cotton and maguey fiber, which survived in dry desert tombs, provide a wealth of knowledge for archaeologists.

Weave as you walk

SANDALS IN THE ANDES
The Incas made sandals with leather from the neck of the llama. In other regions, sandals were made of wool or, as in this case, from the fiber of the aloe plant.

BACK-STRAP LOOM
The most common loom used throughout the Americas was the back-strap loom. It consists of two bars, one hooked to a support, such as a post or tree, and the other to a belt around the weaver's waist. The warp (vertical) threads are pulled taught in between the two. The weft (horizontal) thread is then passed under and over the warp threads using a heddle stick.

BAG MEN
An Inca man would have carried a small bag over the shoulder, under his cloak. It may have contained coca leaves for chewing and amulets (charms).

TO CAP IT ALL
In the Andean region, people usually wore knitted wool or cotton caps. This handsome cap from the coastal Chimu region is unusual because it is made from woven rather than knitted wool.

SOLAR POWER
The most important Inca god was the Sun, Inti, to whom tributes were paid in the hope of improving the harvest. The Sun was a symbol of prestige and power, and the Inca kings believed they were its descendants.

IT'S A WRAP
Andean people mummified corpses, binding them with cord to maintain a seated pose. The mummy was cared for as if it were still alive.

Figure of a god with arms outstretched

ASK MUMMY
Mummies were wrapped in ornate woolen cloths and surrounded by goods that might be useful in the afterlife. They were frequently consulted by the living on important matters.

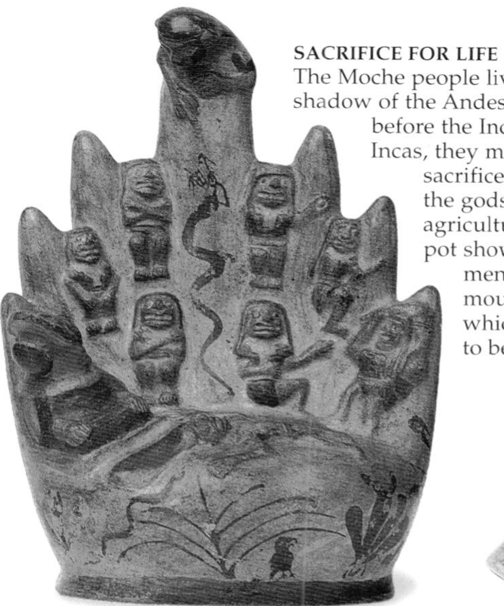

SACRIFICE FOR LIFE
The Moche people lived in the shadow of the Andes centuries before the Incas. Like the Incas, they made human sacrifices to appease the gods of water and agriculture. This Moche pot shows a group of men sitting in the mountain peaks, which were believed to be sacred places.

CARVED KNIFE
Knives of various shapes with metal blades were known as *tumi*. This Peruvian *tumi* is made of copper with a bone handle.

Handle shaped like an animal's head

COOKING POT
This pot was found in an ancient Peruvian grave. It is an elaborate and highly decorated version of the type of pot that would have been used in the Andean region for everyday cooking and eating.

PANPIPES
A common instrument in the Andes was the syrinx or panpipes, usually made of cane or pottery tubes of different lengths. Delicate sounds were produced by blowing across one end of the panpipes. These Inca pipes are made from the quills of a bird of prey called the condor.

Horizontal quill secured with string holds pipes together

North American mountain people

WITH TEMPERATURES DROPPING TO –30°F (–34°C) during harsh winters, the mountains of North America are almost inaccessible, except during the short summer season. The most dramatic peaks are found in the Rocky Mountains, which run from British Columbia in Canada all the way south to New Mexico in the US. A young range, at their highest point the Rockies exceed 13,000 ft (4,000 m). The first people to cross the mountains were native North Americans, who migrated across the high passes. Pioneering settlers arrived later in the search for fur, gold, and other minerals. They made goods from local materials and traded with the native people.

Thompson tribe members, 1925

NATIVE MOUNTAIN DWELLERS
Groups of Native Americans have inhabited the Rocky Mountains for thousands of years. The invasion of the region by Europeans and the conflict that ensued led to widescale migration among some tribes. Many native people now live on protected reservations in mountainous areas.

TRAVEL AND TOURISM
In recent times, the Rocky Mountains have attracted many visitors. The spectacular landscapes surrounding towns such as Aspen in Colorado (above), provide excellent facilities for tourists, including nature reserves in the summer, and ski slopes in the winter months. Other people enjoy fishing in rivers or bathing in hot springs. The construction of major roads over high mountain passes has made Banff national park in Canada and other recreational areas more accessible.

19th-century Blackfoot warrior's shirt

Practice bow and arrow set used by a boy of the Thompson tribe

Wooden practice arrow

Painted symbols

Feathers decorate bow made from mountain sheep horn

HUNTING
Before the arrival of the first European settlers, the mountain slopes offered a rich hunting ground for native people. The men were responsible for making weapons and hunting for game while the women gathered wild plants. The best bows were made from the horn of mountain sheep. Such bows had a long range and good accuracy, and so became important trade items.

PONY EXPRESS
From April 1860 to October 1861, the Pony Express delivered mail between St. Joseph in Missouri and Sacramento in California. The route covered 1,800 miles (2,900 km) and took about 10 days, with riders changing horses as many as eight times.

TRADITIONAL CLOTHES
Native North American people have been wearing traditional garments for centuries. The Blackfoot people made elaborate shirts which were worn by warriors on important ceremonial occasions. Some shirts were made of animal skins and decorated with paint, glass beads, feathers, and even human hair. The most expensive ritual costumes could be traded for up to 30 horses.

Scalp-hair fringes

The Donner monument

THE DONNER PARTY
The Donner monument in the Sierra Nevada mountains of California commemorates the tragedy of the 80-member Donner Party of 1846–47. Led by George and Jacob Donner, the settlers were travelling toward the Sacramento Valley. Stranded by blizzards, their food ran out, and almost half of the group died before the others were rescued.

PURE NUGGETS
Gold is just one example of many minerals that have been extracted in the North American mountains. During the 1890s, prospectors endured bitterly cold winters in their desperate quest to find nuggets of the precious metal.

A golden nugget

A hotel in Dawson City in the Klondike region, 1898

GOLD RUSH
When the first prospectors discovered gold reserves, thousands flocked to the Yukon Territory in northwestern Canada, hoping to make their fortune. The great gold rush of 1898 occurred when rich deposits were found in Bonanza Creek, a tributary of the Klondike River. However, the boom did not last long. By 1900, most miners had already left the area disappointed.

Gods, myths, and legends

Chinese five-clawed dragon

MOUNTAINS ARE mysterious places. Their peaks are often hidden in the clouds, far removed from the comfort of the towns and villages below, so it is no wonder that they inspire awe in all who see them. Throughout the world, this fascination is expressed through a wealth of stories about gods, mythical beings, and weird beasts who inhabit the mountaintops. Most of these legends have been discounted as fantasy or have a scientific explanation. But some, such as the story of the Yeti – a giant apelike creature thought to prowl the mountains of Nepal and Tibet – have defied attempts to disprove them. So perhaps one day a Yeti will be found in the snowy wilderness of the Himalayas.

THE CHINESE DRAGON
In Chinese mythology, mountains are thought to form where dragons coil up beneath the land. The Chinese dragon is the ultimate symbol of good fortune, but in Western mythology dragons can represent chaos or evil that needs to be controlled. In Celtic and European myth, dragons' lairs are often high on mountains.

ANIMALS AND GODS
The ram features in the mythology of Tibet. Mountaintops are the mythical home of many animals and gods, especially those representing rain, the Sun and thunder. In many traditions, the mountain has female characteristics, while the fertilizing sky is male.

PEAK OF PILGRIMAGE
In Japan, Mt. Fuji is a sacred mountain. Climbing to its 12,400-ft (3,776-m) summit has long been a practice of the Shinto religion. In early times pilgrims wore white robes. Today more than 100,000 people a year ascend, most in the climbing season during July and August.

Mt. Fuji has become an emblem of Japan itself

Reconstruction of Aztec temple in Teotihuacan

Stone carving of Tlaloc, the Aztec rain god

GOD OF RAIN
The Aztecs believed that the god Tlaloc governed fertility. He brought both the gentle rain that watered their crops and devastating storms. Not surprisingly, Tlaloc was associated with mountaintops where rain clouds gathered. During his festival, young children were sacrificed on mountain summits to appease the god.

Carved skulls add to the drama of the shaman's dance

In Greek myth, Zeus was born on Mt. Lycaeum and grew up on Mt. Ida in Crete

HOME OF THE GODS
In many ancient cultures, mountains were thought to be places where heaven met Earth. Other beliefs held that mountains were the central axes of the world, the supports of the skies, or the source of life-giving water because rivers often flowed from them. Gods were born and lived on mountaintops. Zeus, the chief of the ancient Greek gods, chose Mt. Olympus (in mythology, the highest peak in the world) as home for the new ruling gods. These gods came to be known as the Olympians.

HOPEFUL HEIGHTS
The Old Testament records that after a great flood, Noah's Ark alighted on Mt. Ararat. In this biblical story, the mountain symbolizes salvation and new life. Similar stories of flood, followed by hope and renewal appear in various cultures throughout the world.

Tibetan masks can depict gods, demons, or even death

Mt. Ararat (or Mt. Agri) lies on the border of Turkey and Iran

TIBETAN MOUNTAIN DANCE
In the Himalayas, shamans (priests) perform dances wearing brightly colored wooden masks. The shaman is believed to take on the attributes of the god or animal that the mask represents. In this way he can influence natural events and ward off evil spirits. The dances, which are performed both on festive occasions and at times of ill fortune, are to a slow rhythm produced by deep-sounding drums.

Homes in high places

ARTISTS IN STONE
The Incas were South America's first great builders. They constructed cities on steep, difficult terrain, and were renowned for their fine stonework. With nothing more than stone hammers and wet sand, they ground irregularly shaped white granite stones to fit together so closely that there was no need for mortar. The results of their craft can be seen at the spectacular city of Machu Picchu and at the Inca capital of Cuzco, both in modern Peru.

IDEALLY BUILDINGS SHOULD BE CONSTRUCTED on level ground where firm foundations can be laid, where access is good, and where the weather is mild. Building in the mountains clearly poses many practical difficulties. Generations of hill dwellers have overcome these problems and, in the process, created distinctive styles of mountain architecture. For example, Alpine chalets have long sloping roofs that hang down over the walls to prevent snow from falling on the doors and windows.

In the high mountains there are few trees, so homes are built of stone. Typically they have thick walls and tiny windows to conserve heat during the cold winters. Many have strong, flat roofs on which snow can build up to give an extra layer of insulation.

Rock cones called fairy chimneys are sculpted by wind and rain

HOUSES OF THE HILLS
In the inhospitable mountains and plateaux of central Turkey, wind erosion has left tall cones of rock standing up where softer stone and soil have been blown away. Some of these structures have been quarried for building stone, and others have been hollowed out to make houses.

Thick thatched roof to keep the weather out and the heat in

THICK THATCH
The first houses ever built were probably covered with thatch. Straw, reeds, and bamboo are still widely used as roofing materials. In mountainous areas the thatch is typically much thicker than in the lowlands, overlapping many times to keep out the extreme weather. In the Nepalese foothills of the Himalayas, a sound roof is essential because of the torrential monsoon rains that fall between June and September. Up to 12 in (30 cm) of rain can fall in one month.

Pierced screens shield the interior of the palace from intense sunlight

Alpine house made of wood

ALPINE RETREAT
Traditionally, Alpine huts were built of wood and had shuttered windows to conserve heat. In the winter months, cattle lived on the ground floor and people lived upstairs. Today similar huts are scattered across the Alps, providing shelter for climbers, walkers, and skiers. Often they are sited in high "eagle nest" positions that can be spotted from far away and are clear of potential avalanches.

LOFTY HIDEAWAYS
People build on high, rocky outcrops for three main reasons. The first is defense – an elevated position makes it easier to see enemies approaching. The second is tradition or belief – mountains are often thought to be sacred places. The third reason is status – a building on high reflects the social status of its owner. Shown here is the Summer Palace at Wadi Dahr in Yemen, home of the powerful Imam Yachiya.

Making a living

SPINNING A YARN
Few people left a greater wealth of woven textiles than the ancient Peruvians. They either used cotton thread (seen here) or the wool of alpacas and llamas.

THE RUGGED SLOPES of high mountains appear barren and unproductive. The climate is cold, and growing crops on the poor, thin soils is very difficult. But mountains can be a source of great wealth. Some are mined for gold, silver, or other precious metals and minerals. Others are exploited by loggers who cut large amounts of timber. As coal and oil become more scarce, hydroelectric power stations that harness the force of mountain rivers will become ever more important.

STEPS FOR SURVIVAL
Building terraces is a practice that goes back thousands of years. It allows crops such as rice, corn, vegetables, and fruit to be grown on steep slopes. Here, retaining walls create flat pockets of land and hold back the soil and water that would otherwise wash down the hillside.

HUNTING FIREWOOD
In parts of the Himalayas, so much of the forest has been cut down that people have to travel for days to collect enough firewood to cook their food and heat their homes.

Industrial strength

Mountain resources are key to many heavy industries. Steel mills in the Austrian Alps, for example, make use of rich deposits of iron and coal in the mountains, and paper mills throughout the world are often found in the hills, where timber is plentiful.

ENERGY FROM WATER
Mountains appear to be ideal places to build hydroelectric power stations. Hilly terrain and high rainfall mean that large volumes of water can be trapped behind tall dams. The stored water is then allowed to fall through turbines that produce electricity. Mountainous countries, including Norway, Sweden, Canada, and Switzerland rely heavily on this type of electricity generation.

Concrete dam

CRYSTAL CLEAR
Quartz crystals have been collected in huge quantities from the Swiss Alps for at least 200 years. The large specimens are sought after by collectors, while the smaller ones are used in the optical and electronics industries.

Gwindel quartz from the Alps

Molten gold was poured into a mold

Gold Inca llama

PANNING FOR GOLD
The Incas obtained most of their gold from "placer" mines in rivers, where the gold is near the surface. They used fire-hardened sticks to break up the gold-bearing soil.

TOUGH FOOD
High in the mountains, the growing season is short and the winters cold. Only the hardiest crops, such as beans, potatoes, and barley, are planted.

Beans grown in the Andes

Corn is a staple food in many mountain regions

LIVING OFF THE LAND
Nepal is one of the poorest countries in the world. Many of its inhabitants survive by subsistence farming, which means that they use all the crops they grow, with nothing left over to sell. For them, using agricultural machinery is out of the question. The land must be plowed, and crops cultivated, by hand.

UP IN THE AIR
After the industrial revolution, efficient rail links brought tourists to the Alps for the first time. Visitors enjoyed rides on aerial lifts.

Mountain travel

STEEP TERRAIN HAS ALWAYS BEEN a barrier to communication. Early mountain travelers had no choice but to walk along well-established passes, and as long ago as 3000 BC large settlements sprang up where these passes converged. With industrialization came the railroads, and steam engines were soon adapted to work in the mountain environment. Today, the old mountain passes have been expanded into highways. Huge bridges and tunnel systems, such as the 12.3-mile (19.8-km) long Simplon II tunnel in Switzerland, carry roads and railways into once remote valleys. Good transportation networks help develop mountain economies, but can also bring environmental damage to these sensitive areas.

Traditional snowshoe has a light, wooden tennis racket-shaped frame and is strung with thongs

LIGHT FOOTED
Snowshoes were first used by Inuit people in the cold northern parts of North America. They spread the wearer's weight over a large area, allowing him or her to walk on soft snow without sinking. Today, showshoes are made from aluminum and snow-shedding plastics.

Train to the summit of Mt. Snowdon, 3,560 ft (1,085 m) above sea level

TRACKS TO THE TOP
Trains cannot climb steep slopes, so track usually has to be laid in shallow spirals that wind up a mountainside. However, some trains, including the one that carries visitors up Mt. Snowdon in Wales, UK, work in a different way. A giant cog drives the train up a toothed track, preventing it from rolling back. Such cog-wheel trains can move upward on slopes of nearly 50° inclination.

Backpack holds essential equipment for survival in the wilderness

Ski touring is growing in popularity because it is an excellent way of visiting the mountain environment and keeping in shape

CROSS-COUNTRY SKIING
Skiing as a form of transportation is more than 4,000 years old. Historians believe it began when people living in Scandinavia strapped the bones of large animals to shoes using leather thongs. Today people tour the mountains on skis for fun. "Skins" attached to the bottom of the skis provide friction so that skiers can travel uphill as well as downhill.

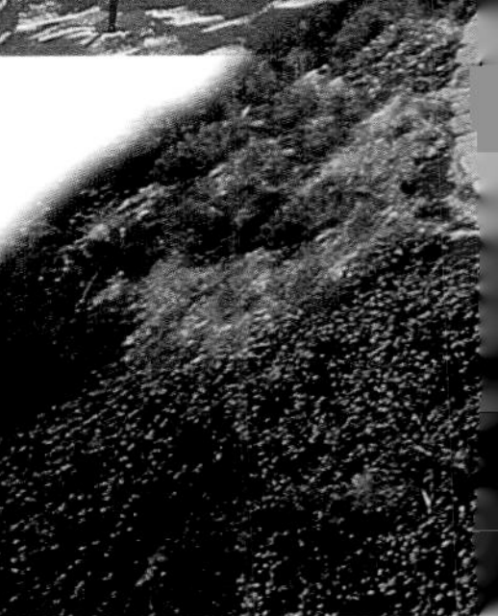

A BRIDGE TOO FAR
Suspension bridges stretched across river valleys are a common sight in mountainous regions around the world. Whether they are made of simple wood and rope, or metal wire and concrete, they offer easy access across gorges for travelers and pack animals.

CABLE CAR
Cable cars may be used for sightseeing, or for transporting mine workers into mountainous environments. Most often they carry skiers to the tops of mountains. This cable car in Palm Springs, California, carries tourists to a height of 8,516 ft (2,600 m) in less than 15 minutes.

Each car can hold 80 passengers

TAKING THE STRAIN
Pack animals can carry heavy loads over the roughest mountain terrain. Ideally, they should be hardy with a calm temperament. Donkeys, mules, llamas, camels, and yaks are all used in different parts of the world.

LET'S TWIST
Roads are often cut into steep mountains in a zigzag pattern, which produces perilous hairpin turns but reduces the gradient that cars must negotiate. While roads have ended the isolation of mountain communities, they have also brought air and noise pollution to busy Alpine valleys.

Hairpin turn

Early days of mountaineering

MOUNTAIN SUMMITS HAVE attracted people for thousands of years. For ancient people, they were the home of the gods, and arduous ascents were made to leave offerings for the deities or to communicate with the spirit world. By the 18th century, scientific study of mountain features, such as glaciers, plants, and animals was taking naturalists high into the European Alps, while prospectors were drawn to the mountains in search of precious stones and metals. The era of mountain exploration had begun, and mountaineering as a sport soon followed. Groups of British climbers assisted by expert French, Swiss, and Italian guides, had scaled all the main Alpine summits by the 1870s.

ARMY OF CLIMBERS
One of the most remarkable of all mountain journeys was made more than 2,000 years ago. Hannibal, shown on this coin, was a great general from the Carthaginian province of Spain. In 218 BC he led a force of more than 9,000 men and 57 elephants across the rugged, snowy slopes of the Alps to launch a surprise attack on the Romans.

CANINE FEATS
Perhaps the most unusual of the early mountaineers was Tschingel the dog, who achieved an impressive list of Alpine ascents, accompanied by his American mistress Meta Brevoot. Tschingel achieved such fame in Switzerland that a mountain was named after him.

Triumph at the summit

"I could not look upon the mountain... without being seized by an aching desire."

HORACE BÉNÉDICT DE SAUSSURE

WOMEN ON HIGH
The achievements of the first women climbers of the late 19th century deserve special credit. As well as venturing into what was considered a man's world, often fighting the disapproval of friends and family, women were obliged to climb wearing cumbersome skirts.

Guide

The fan commemorates Smith's expedition

The reverse of the fan served as a program for Smith's lecture

SOUVENIRS OF ADVENTURE

The Englishman Albert Smith wasn't the first to climb Mont Blanc. He reached the summit in 1851, 65 years after the first ascent of the mountain, but his showmanship helped boost interest in mountaineering. Smith staged a spectacular illustrated lecture called *The Ascent of Mont Blanc* in London's Piccadilly and produced souvenirs, such as a board game and this fan, which fueled public fascination for the new sport of climbing.

MONT BLANC CHALLENGE

Scaling the Alpine peaks became an obsession for 18th-century adventurers. In 1760, the Swiss physicist Horace Bénédict de Saussure offered a prize for the first ascent of Mont Blanc, which at 15,770 ft (4,807 m) was the highest peak in the Alps. The prize was claimed 26 years later by Michel-Gabriel Paccard, a French doctor, and Jacques Balmat. De Saussure made his own ascent the following year accompanied by 18 guides and his personal valet.

Whymper plants a flag on the summit

Whymper's party at the summit of the Matterhorn

TRAGEDY ON THE MATTERHORN

By the mid-1800s, one Alpine peak – the Matterhorn, on the border between Switzerland and Italy – still seemed impossible to climb. An English artist called Edward Whymper (1840–1911) was determined to be the first to reach the summit of the 14,692 ft (4,478 m) mountain. In 1865, on his eighth attempt, he finally succeeded. But the expedition ended in disaster. On the descent one of his companions slipped and fell, and four people plunged to their deaths.

THE ICEMAN

The first known mountaineer, remarkably preserved in a glacier at an altitude of 10,531ft (3,210 m), is Ötzi the Iceman. Discovered in 1991 in the Ötz valley between Austria and Italy, his mummified body was first thought to be that of a modern climber, but carbon dating showed his true age to be 5,200 years. Ötzi's clothes, shoes, and tools showed that he was well used to life at high altitude.

Ötzi's wrist bears tattoos, once thought to have healing qualities

The body was packed in snow and ice

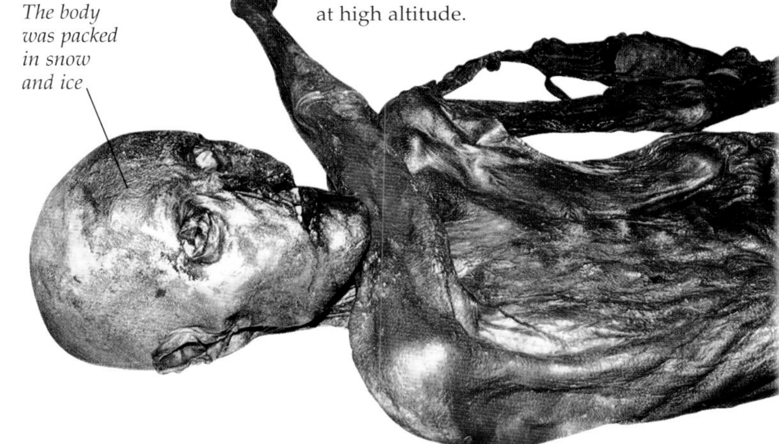

The ultimate challenge

MARKING DEFEAT
Fellow climbers waited for news of Mallory and Irvine at a lower camp. They saw members of the party walking away from a cross they had made by laying blankets on the snow. This signal spelled "death." Mallory and Irvine were lost.

AS MORE AND MORE peaks in the European Alps were climbed, mountaineers set their sights on summits in other parts of the world. The biggest prize of all was Mt. Everest. Reaching its peak would be a journey into the unknown, because before 1921, no climber had gone above 24,600 ft (7,498 m) on any mountain – 4,429 ft (1,350 m) short of Everest's summit. In 1921, a British expedition rose to the challenge and discovered a possible route. In 1922, they returned and climbed to within 1,700 ft (518 m) of the summit. In 1924 they returned again – and tragedy struck. On June 8, George Mallory and Andrew "Sandy" Irvine vanished, never to be seen again. To this day, no one knows if they made it to the top.

EVEREST – MAN AND MOUNTAIN
Mt. Everest is named after Sir George Everest, the son of a London lawyer, who was Surveyor General of India from 1830 to 1843. As a tribute to his work in mapping India, his successor Andrew Waugh proposed naming the world's highest mountain after him. Everest didn't like the idea, believing that mountains should be known by their local names. Waugh won the argument, and the name remains.

Andrew "Sandy" Irvine
George Mallory
E.C. Shebbeare
Colonel "Teddy" Norton
Noel Odell
John Macdonald
Bentley Beetham
Howard Somervell
Geoffrey Bruce

TEAM OF 1924
Under the leadership of Colonel Norton, the 1924 team included a nucleus of climbers who had been to Everest before – Somervell, Bruce, and, on his third Everest expedition, George Mallory. Newcomers included Odell, Beetham, Hazard, Shebbeare, and Hingston. Young Sandy Irvine, an Oxford engineering student, aged only 21, had little climbing experience but was very strong and, it is said, always cheerful.

THE MISSING PHOTOGRAPH
Just before Mallory set off for the summit he was given a Kodak Vestpocket camera. The likelihood is that he or Irvine carried it on that fateful day in June, and that the camera lies hidden somewhere high on Everest to this day. Experts believe that if it were found, the film could be developed and answer once and for all the question – were Mallory and Irvine the first to reach the summit of Everest?

The camera folds down to fit into a pocket

Mallory is wearing a windproof jacket, breeches, and cashmere puttees

Oxygen cylinder

Each climber carried a few spare clothes and provisions

FINAL PUSH TO THE SUMMIT

The last picture ever taken of Mallory (left) and Irvine alive shows them at Camp 4 on Everest's North Col, making final preparations to climb to the peak. They carry on their backs heavy oxygen cylinders, which their Sherpas and local Tibetans jokingly said contained "English air." Noel Odell was the last man to see the two alive, glimpsing them high on the North Ridge at 12.50 pm, still going strong toward the summit.

The note was scribbled on a page torn from Mallory's notebook

LETTER TO NOEL

Mallory's note, written at high camp 6, tells Captain John Noel, where he might spot the two on their final approach to the summit. It reads: "We'll probably start early to-morrow (8th) in order to have clear weather. It won't be too early to start looking for us either crossing the rock band under the pyramid or going up skyline at 8.0 pm [he meant 8 am]."

Irvine's ice-ax

> *"Because it's there"*
>
> MALLORY'S FAMOUS ANSWER TO THE QUESTION OF WHY HE WANTED TO CLIMB EVEREST

Climbing rope

> *"...the whole fascinating vision vanished, enveloping in cloud once more."*
>
> NOEL ODELL ON HIS LAST SIGHTING OF MALLORY AND IRVINE

FROZEN EVIDENCE

The first clue about Mallory and Irvine's fate was an ice-ax found high on the mountain in 1933. The bodies of the two men were nowhere to be seen. But remarkably, in 1999, 75 years after they disappeared, the body of Mallory was found. His injuries suggested that he had fallen, but there is not enough evidence to prove whether or not he and Irvine made it to the top.

Hillary and Tenzing

UNTIL WORLD WAR II, all efforts to reach the summit of Everest were made by British expeditions. All of them climbed from Tibet in the north, because the southern approach from Nepal was closed. After the war Nepal opened its borders, and new maps, clothing, and equipment became available, allowing the British to pioneer a route from the south. In 1952, a Swiss expedition used this route to climb within 1,000 ft (300 m) of the summit, and the race for Everest was on. The following year, the British set forth again, determined to conquer "their" mountain before the Swiss or the French beat them to it.

NEW ROUTE TO THE TOP
This illustration from a contemporary edition of the *Daily Express* shows the path taken by Hillary and Tenzing on their historic climb. Once past the initial obstacle of the Khumbu Icefall (see below right), the route continued along a high valley, and up a steep face toward an exposed ridge. The Swiss expedition of 1952 had climbed to an altitude of 28,200 ft (8,600 m), but above this the terrain was unknown.

Aged 49, Dawa Thondup was a veteran Himalayan climber

Da Tenzing was respected by everyone for his great courage

Ang Tsering was just 16 years old

TEAM OF 1953
The expedition leader was an army officer named John Hunt, a man with climbing experience in the Himalayas and Alps, and a genius for organization. Hunt chose the best climbers from the British Commonwealth. It was his intention that three pairs of climbers would try for the top. The first pair, who made a valiant attempt, but stopped just short of the summit, were Tom Bourdillon and Charles Evans.

EVEREST SCALED
At 11.30 a.m. on May 29, 1953, Sherpa Tenzing Norgay and Edmund Hillary, a beekeeper from New Zealand, became the first men to step onto the summit of Everest. Tenzing posed for this photograph – his ice-axe adorned with the flags of Britain, Nepal, the United Nations, and India – before making a Buddhist offering of chocolate and cookies to the gods of the mountain.

Porters carried loads of up to 67 lb (30 kg)

THE ARMY BEHIND THE CLIMB

Only two men may have stood on the summit of Everest, but an army moved in support of the expedition. A team of 350 porters, half of whom were women, carried equipment and supplies through the foothills of the Himalayas. Around 30 Sherpas – or "Tigers" as they were affectionately known – carried supplies to the high camps on the mountain.

Climbers breathed gas through a mask covering the nose and mouth

Oxygen was contained in alloy cylinders

Loaded with two full oxygen bottles, the backpack weighed a hefty 30 lb (14 kg)

Oxygen set as used by Hillary and Tenzing

BREATHING APPARATUS

The closed-circuit oxygen apparatus used on the ascent was designed by expedition member Tom Bourdillon and his father. While oxygen is cumbersome to carry, it helps to keep the body warm and reduces headaches. For the first time in 1953, it was used as an aid to sleeping. "But for oxygen," wrote Hunt, "we should certainly not have got to the top."

CROSSING THE ICEFALLS

The Khumbu Icefall is a terrifying, tumbled labyrinth of icy rubble, some 3,000 ft (1,000 m) from top to bottom. Climbers on the 1953 expedition took alarming risks and braved fragile snow bridges and huge, unstable blocks of ice as large as houses. But after several days they found a way through the icefalls, and over the following weeks made the route safe so that supplies and equipment could be carried high onto the mountain.

The route was made as safe as possible by using fixed ropes and ladders

DAILY EXPRESS

BE PROUD OF BRITAIN ON THIS DAY, CORONATION DAY

ALL THIS—AND EVEREST TOO!

Crowds singing in the rain

Three girls made the Queen's dream dress

EXCLUSIVE!

BRITON FIRST ON ROOF OF THE WORLD

20th CENTURY ELIZABETHAN

DEVON CYDER

WHITEWAY'S DEVON

HEADLINE NEWS

When the summit was reached, James Morris, a reporter from *The Times*, sent a coded message to London. "Snow conditions bad," he wrote, meaning success! By June 2, 1953, the whole world had heard the good news – just in time to celebrate the coronation of the new British queen, Elizabeth II.

> *"We shook hands and then Tenzing threw his arms round my shoulders and we thumped each other on the back until we were almost breathless."*
>
> EDMUND HILLARY
> ON REACHING EVEREST'S SUMMIT

BACK TO BASE

Hillary (left) and Tenzing celebrate their achievement with a cup of tea. When expedition leader John Hunt saw the climbers descending from the mountain, he took their wearied postures to be a sign of failure. Spotting him, the men raised their ice-axes, and pointed toward the summit. As they did so, Hunt realized the wonderful truth. "Far from failure," he later wrote, "this was IT. They had made it!"

Traverses and triumphs

ONCE EVEREST AND ALL THE 8,000 m (26,247 ft) peaks had been scaled, mountaineers looked for new climbs in remote parts of the world and more challenging ways to climb the same summits. New routes and techniques now test a climber's skill and endurance. Over the years, the Himalayan giants were climbed by huge expeditions, relying on the support of Sherpas and the security of tented camps and ropes fixed to the mountain. Modern climbers attack these mountains in the same way they would smaller peaks in the Alps, in small groups, moving quickly and carrying their own equipment. They want to achieve more with less, and one man who has taken this "Alpine style" to the extreme is the Italian climber, Reinhold Messner.

HIGH-WIRE ACROBAT
Catherine Destivelle is one of most famous climbers in the world. Born in 1960, her love of climbing was kindled at an early age near her home in Paris. She made front-page news when she became the first woman to climb the Eiger and pioneered a new solo route on the Petit Dru in the Mont Blanc range. Films of her acrobatic climbing style have helped popularize the sport of climbing.

Destivelle on the Petit Dru, one of the hardest of all Alpine climbs

MUMMERY THE PIONEER
Many consider the British climber Albert Mummery (1855–95) to have been the founder of the "new route" school of mountaineering. He was a bold climber, even reckless at times, but technically superb, and one of the first to climb without a guide. He made the first ascent of the Zmutt Ridge on the Matterhorn and the north face of the Grepon, both in the Alps. He died with two Gurkha companions while attempting to scale Nanga Parbat in the Himalayas.

UP THE NOSE
The Nose of El Capitan, a formidable lump of rock in Yosemite Valley, California, was first climbed in 1958 by Americans Warren Harding, Wayne Merry, and George Whitmore. They were on the face for 37 days, laboriously hauling all their own food and provisions with them. Canadian Peter Croft has since climbed The Nose alone in a remarkable 4 hours, 22 minutes.

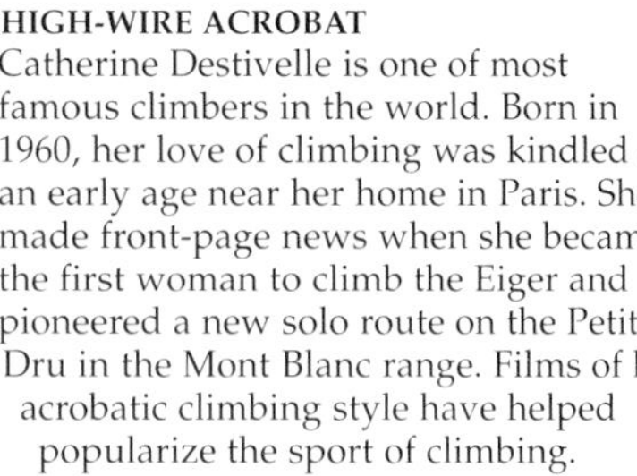

FIRST WOMAN ON EVEREST
In 1975, Junko Tabei, a 35-year-old working mother from Japan, became the first woman to reach the summit of Mt. Everest. She said of her climb: "Technique and ability alone do not get you to the top... it is the willpower that is the most important. This willpower you cannot buy with money or be given by others – it rises from your heart." Only 11 days after her ascent she was followed by Phantog, a Tibetan woman, who climbed Everest from the north. Tabei was also the first woman to climb the Seven Summits – the highest peaks on the seven continents.

SHEER TRIUMPH
Few expeditions have achieved such remarkable success as the 1970 British expedition to Annapurna – the 26,545 ft (8,091 m) peak in the Himalayas. Led by Chris Bonington, the expedition aimed to make the first ascent of Annapurna's South Face, a 12,000 ft (3,600 m) wall of ice and rock that soars almost vertically to the summit. They experienced dreadful weather and had to battle against severe winds for most of the way; but on May 21, Don Whillans and Dougal Haston, unroped and without supplementary oxygen, stood on the summit.

Peter Habeler took up climbing when he was just six years old

Reinhold Messner has inspired a generation of climbers

THE ULTIMATE ALPINIST
Italian Reinhold Messner (left) and Austrian Peter Habeler (right) made history in 1978 when they became the first men to climb Everest without the aid of bottled oxygen – a feat that few thought possible. Two years later Messner not only repeated the ascent without oxygen, but did so alone. Messner was the first to climb all the fourteen peaks in the world over 8,000 m (26,247 ft), and is considered by many as the greatest Himalayan climber of all time. A true climbing purist, Messner considers aids, such as bolts and oxygen in tanks, to be "unfair means" to scale a mountain.

SOLO HERO
The Italian Walter Bonatti is one of the most accomplished climbers of all time. This picture from 1965 shows him at the top of the North Face of the Matterhorn having just become the first person to climb it alone, and in the middle of winter. His most remarkable ascent was an epic six-day solo climb in 1955 of an impossibly steep column – the South West Pillar of the Dru – in the Mont Blanc range. It is known today as the Bonatti Pillar. Bonatti said: "If in normal conditions, it is skill which counts; in such extreme situations, it is the spirit which saves."

Clothes and equipment

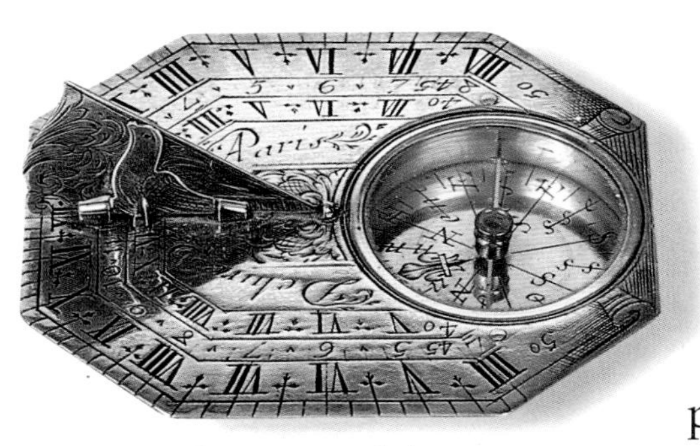

17th-century sundial compass

THE FIRST MOUNTAINEERS set off into the hills with only the most basic of equipment. A long, pointed walking stick, or alpenstock, helped with balance, while crampons gave a surer footing on ice, and axes were used to cut steps into the ice. Clothes were made of wool, cotton, and silk, which gave warmth when worn in layers, but became sodden in the wet. Today's climbers are armed with a variety of sophisticated garments and gadgets developed specially for use on the mountain. These make life at high altitude safer and far more comfortable, but also allow people to venture into steeper, colder, and more hostile environments. Clothes made from artificial fibers are warm and waterproof, and the use of materials including titanium, aluminum, and carbon fiber means that vital equipment, such as ice-axes, can be stronger and lighter than ever before.

Hat to keep off the strong Alpine sun

Detachable skirt worn for respectability

SKIRTS ON THE SLOPES
In an age when every woman wore a skirt, Mrs. Aubrey le Blonde broke convention by climbing in trousers. A skirt worn over the top could be removed as soon as she was out of view of inquisitive villagers. Once she left her skirt on a mountain top and had to climb back up again to retrieve it before entering the respectable inns in the valley.

Traditional ice-axes were about 3 ft (1 m) long with a spike at one end and a pick and hammer at the other

Alpenstock

Ice-ax

ALPENSTOCK AND ICE-AXE
Long, pointed walking sticks called alpenstocks (from the German for "Alp sticks"), and axes for cutting steps in the ice were vital tools for 19th-century climbers. These were superseded by the ice-axe which served the functions of both. The first ice-axes were made of bamboo or hickory.

Primitive crampon made from wood and rope

SNOW SUPPORTS
In the 16th century shepherds high on Alpine slopes would strap simple wooden frames with downward-pointing pegs to their boots to help them walk safely over snow. The idea was borrowed in the early 1900s to make the first metal climbing crampons.

Leather hobnail climbing boot from the 1920s

BOOTING UP
The early mountaineers climbed in hobnail boots. Nails embedded in the sole give a good grip on granite, but on snow and ice a frame of metal spikes called a crampon was strapped to the boot for surer footing. Modern crampons are stamped from lightweight steel and have two front-pointing spikes that can be dug into steep ice walls.

Leather straps secure the crampon to the boot

This geodesic tent is lightweight but offers plenty of space for two people

DOMED TENT
A portable shelter has always been an essential piece of mountaineering equipment. Traditionally tents were A-shaped with wide, flat sides. They were heavy to carry and prone to being knocked over by strong winds. Today, tents are made in the shape of a geodesic dome, or are rectangular with a curved roof. They are stable in winds of up to 80 mph (130 km/h) and can withstand heavy snow. Where older tents had two separate layers, modern lightweight designs have just one, usually made from a "breathable" fabric, such as Gore-Tex.

A drawstring around the neck ensures that a climber's body heat stays in the sleeping bag

Tapering shape makes best use of body heat

SNUG AS A BUG
Sleeping bags keep climbers warm by trapping air between strands of a filling material, which may be goose or duck down, or a synthetic fiber. Most mountaineers prefer the cosiness of down filling, although it does not perform so well in wet conditions. Some sleeping bags have a waterproof shell, or they can be slipped inside a waterproof bivouac bag if dampness is likely to be a problem.

A helmet gives protection from falling rocks

Snow goggles prevent snow-blindness, a condition where the eyes are "sunburned" by intense ultraviolet light

A waterproof jacket made of a breathable fabric keeps off rain and snow

Climbing ropes are slightly elastic to break a climber's fall

The jagged stainless steel head of an ice-ax grips ice and snow

Waterproof mitts are worn over fleece, wool, or down gloves

Waterproof shell worn over long johns and fleece trousers

DRESSED FOR ACTION
A modern high-altitude climber wears three main layers of clothing. Thermal underwear made of nylon helps draw moisture away from the skin. A thicker layer of wool or polypropylene provides insulation. The outer suit provides protection against windchill, snow, and rain.

Gaiters keep the snow out of a climber's boots

Insulated plastic boots have rigid soles so that crampons can be fixed securely

Surviving extremes

VENTURING UP THE WORLD'S highest peaks is taxing for both body and spirit. Wild weather is to be expected, and climbers risk injury from falls, avalanches, and rockslides. As the human body is stretched to its limits, physical problems may set in, such as altitude sickness, snow blindness, and frostbite – when flesh freezes and turns numb and gray. Modern technology helps to minimize the danger – satellite telephones relay weather forecasts, and high-tech equipment improves safety margins. Despite these aids, there is no substitute for a mountaineer's skill and experience.

EFFECTS OF ALTITUDE
Early signs of altitude sickness are breathlessness and a fast pulse, but these can be quickly followed by nausea, headaches, muddled thinking, dizziness, and eventually unconsciousness. The body needs time to adapt to the thin air (air low in oxygen), so it is not wise to climb too quickly! The rule of thumb is to ascend no more than 3,000 ft (1,000 m) a day.

IN A HOLE
In the most extreme conditions, digging and sheltering in a snow hole can be safer than in a tent. Thick snow walls provide good insulation from the cold and protection from blizzards. A hole is dug with an ice-ax or shovel and the entrance is sealed with snow blocks.

THE BREATH OF LIFE

On the summit of Everest the amount of oxygen in the air is one third that at sea level. Few people can climb to such a great height without breathing oxygen from cylinders. Here Britain's Chris Bonington and Sherpa Ang Lhakpa can be seen wearing full oxygen apparatus on their 1985 expedition.

Oxygen mask

Strong, waterproof padded bag protects medical supplies

Sterile dressings

EMERGENCY AID

The upper slopes of a mountain are far away from medical facilities. Past "siege" expeditions often took a medical doctor on the climb, but today's parties must at least carry a good first-aid pack. Typically this contains bandages and dressings, antiseptics, painkillers, antibiotics, drugs for nausea, indigestion, diarrhea, respiratory problems, eye and ear infections, and other emergency medical supplies.

COLD LIGHT

Snapping a chemical light stick provides bright light for up to 12 hours in an emergency if flashlight batteries have failed.

Mountain rescue

By their nature, mountains are dangerous places and accidents do happen, even on the best prepared expeditions. If a member of a party is injured or buried in an avalanche, there may not be time to wait for help, so self-reliance is vital. The first action is always to make sure that the casualty and the rest of the party are clear of further danger. First aid is then given and assistance summoned by radio or phone, or using the internationally recognized distress signal of six blasts on a whistle or six flashes of a flashlight. In most mountain regions there are rescue teams who travel on foot, skis, or by helicopter to people's aid. Their stories can be truly heroic.

MOUNTAIN RESCUE COMMITTEE
RESCUE POST

SIGNS OF HELP
Some countries have professional mountain rescue teams; others have dedicated volunteers who are on stand-by 24 hours a day, 365 days a year. In the British Isles, the Mountain Rescue Council has set up a series of first-aid posts in the hills where emergency equipment is stored.

Sleeping bag or blankets to keep the victim warm

Light alloy frame to ease the load

SNOW STRETCHER
An all-too-common sight on the ski slopes, the blood wagon is effectively a stretcher on a sled which allows an injured person to be carried to safety over difficult and snowy terrain. Poles protruding from each end allow the rescuers to hold it securely while they negotiate the slopes on skis or in climbing boots.

Crampons give a secure footing on steep, icy ground

ON THE SCENT
Dogs have a keen sense of smell and are used to sniff out victims lost in the mountains or buried in the snow. This traditional image of a St. Bernard dispensing brandy from a barrel around its neck is misleading – drinking alcohol causes the body to lose heat and should be avoided in the cold.

Rescue pack

A rescue team sets out into the mountains fully prepared for any eventuality, their backpacks brimming with as much useful equipment as they can carry without overburdening and slowing down the party. A climber or skier venturing into the wilds might carry much of the same equipment.

AIR ASSISTANCE
Sometimes the only way to carry a person to safety from the inhospitable terrain of a mountain is by helicopter. Air crews show extraordinary skill in plucking injured parties off steep cliffs and from deep gullies. Helicopter rescue is impossible on the highest peaks, where the air is too thin to support flight.

Climbing techniques

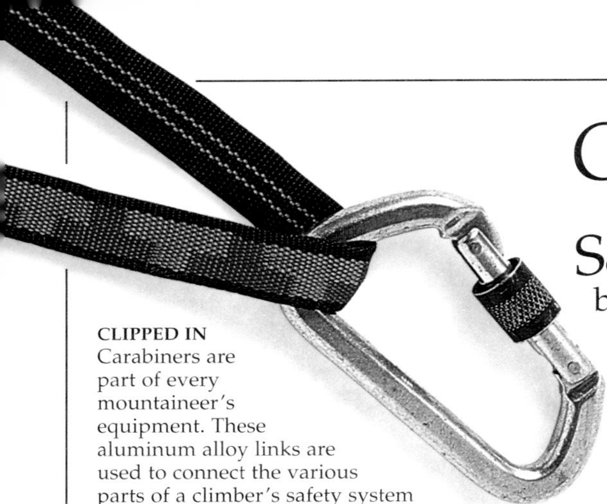

SCALING ANY PEAK demands excellent coordination of eyes, brain, hands, and feet. But mountains are very diverse, each one demanding a different range of climbing techniques and qualities in the mountaineer. At the very highest altitudes, just walking can be hard enough, so determination and physical strength are paramount, whereas climbing rocks, snow, and ice requires skill in handling rope.

Good climbers are always fit, supple, and have a catlike sense of balance. The most committed climbers train hard, pushing their bodies to the limits. This allows them to scale vertical and even overhanging ice cliffs and rock walls in balletic fashion.

CLIPPED IN
Carabiners are part of every mountaineer's equipment. These aluminum alloy links are used to connect the various parts of a climber's safety system – the rope, anchor, belay, and harness. The strongest can support weights of 5,000 lb (2,250 kg).

FALL STOP
Ropes fixed to rock provide the best security on the mountain. But even when unroped, a climber can prevent a fall by throwing his weight onto his ice-ax and driving it into the snow or ice. Sometimes climbers are roped together for safety – if one falls from a ridge, then the others must jump over the opposite side to save everybody's skin.

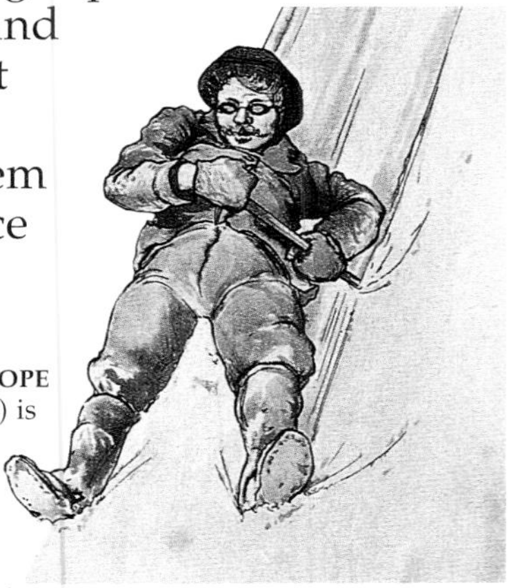

SLIPPERY SLOPE
Glissading (sliding) is a quick way down a snow slope. An ice-ax acts as a brake.

BELAY RELAY
To climb a pitch (rope-length) of rock or ice, the lead climber ascends and fixes the rope to a rocky projection or metal anchor. Then he threads it through a belay plate, or safety catch, secured to a waist harness. This is called belaying, and it allows the lead climber to apply the brakes if his partner falls.

Lead climber anchored securely to the rock

The belayer guides the rope

A pitch – the length between two belays

Second climber secured by the anchor and belay above

FINGERS IN A JAM
Sometimes the only way to get a grip on a rock face is to force a hand or a finger into a crack. Twisting or flexing the hand makes a wedge that locks tightly in the crack. Such finger or hand jams are strenuous and painful. Often they cause cuts, scrapes, and scars, but seasoned mountaineers know that the ability to make a good hand jam is the skill that makes or breaks a climber.

Hand twisted to provide purchase on the bare rock

A pocket for foot or hand holds

INDOOR ADVENTURES
Indoor climbing walls made of brick or resin, molded to resemble rock, provide a place for beginners to learn climbing skills in safety, and are excellent training grounds for the real thing. Climbing on walls has emerged as a gymnastic sport in its own right – many of its practitioners never climb on a real mountain!

COOL CLIMBS
Up to the 1960s, climbers negotiated steep walls of ice by cutting steps – small footholds – using their ice-axes. Today, mountaineers use a different technique, kicking the forward-pointing spikes of their crampons into the ice to support their feet, and using an ice-ax in each hand to claw their way up the wall. Ropes are anchored to surrounding rock or to titanium ice screws (which can be up to 12 in/30 cm long) embedded in the ice itself.

Jagged edge of ice-ax gives good grip

Curved pick of ice-ax

Spade-shaped adze of ice-ax for chopping ice

Climbing harness

Crampons with front-pointing spikes attached to rigid boots

Mountain sports

MANY PEOPLE ARE ATTRACTED TO mountains to climb, but others are drawn by the natural beauty of the landscape, to walk, or to enjoy one of a growing number of mountain sports. Skiing, the most popular of these activities, has ancient origins. Skis have been used to travel in the mountains for at least 4,000 years, and skiing developed into a competitive sport as far back as the 1840s. Today, skiing is a giant industry with custom-built resorts and dedicated transportation networks. Those looking for a challenge go beyond the manicured slopes, or pistes, and ski "extreme" down steep faces and gullies, or tour in untamed valleys. Others try their hand at snow-boarding, mountain-biking, or canyoning.

DANGEROUS DESCENTS
Canyoning is a relatively new sport. It involves traveling down a canyon or a gorge from top to bottom. This might include climbing, rappelling down a waterfall, or leaping from a great height into a rock pool. Like many of the new mountain activities, it is only for the intrepid!

BIG BUSINESS
Today, ski resorts worldwide compete with each other to attract visitors. Ski-tourism is a major source of income for many countries.

WHEELS ON THE SLOPES
Mountain bikes are light, strong machines with rugged tires and many gears for riding on rough terrain up and down mountains. Mountain bikes were first built and used by enthusiasts in California in the 1970s. The latest bikes use aerospace technology to minimize weight without sacrificing strength. Their frames, made from Optimum Compaction, Low Void Carbon Fiber, can weigh as little as 2.8 lb (1.2 kg).

Frame can withstand high levels of stress and impact

FUN IN THE SNOW
Many people are introduced to winter sports through tobogganing, either on a sled, or on an old tray stolen from the kitchen. The most important thing to remember is to jump off before crashing into an obstacle!

GOING FLAT OUT
Cross-country skiing involves traveling across undulating or even flat land. It is very strenuous exercise, a good way of touring the snowy countryside, and is also an Olympic sport. The skis above date from the early 20th century – they are 8 ft (2.5 m) long, wooden, and very heavy. Today's cross-country skis are still long and narrow, but made of fiberglass and are extremely light. Only the toe of the shoe is fixed to a binding on the ski, so that the heel can be lifted to push the skier forward.

CONTROLLED DESCENT
Rappelling, also called abseiling, involves sliding down a cliff using a rope and some sort of brake system to slow descent. It is a common maneuver in rock climbing, but has also become an exciting mountain activity in itself. Although easy to learn, abseiling can be dangerous, and rigorous safety measures are needed.

Harness secures the rappeller to the rope

This free-ride board is wide and flexible

BOARD GAMES
Snowboarding began in the 1960s with the invention of the "snurfer," a simple board with a rope handle attached to its nose. It has since grown into a sport that challenges skiing in its popularity, partly because it is quicker to learn and master. With its gravity-defying tricks and spins, and long carving turns, it is a free and exhilarating sport. In 1998, snowboarding made its debut at the Winter Olympics in Nagano, Japan.

The Winter Olympics

THE FIRST WINTER OLYMPIC GAMES took place in Chamonix, France, in 1924. They were dominated by the Scandinavian countries, which scooped 28 of the 43 medals awarded. Since then – except during World War II (1939–45) – the Games have been held every four years. All sports take place on snow and ice, with men and women from all over the world competing in various forms of skiing – including downhill, slalom, and freestyle – as well as ice-skating, ice hockey, bobsledding, and the luge. The scope of the Winter Games has increased over the years – in the 1998 Games in Nagano, Japan, curling, snowboarding, and women's ice hockey made their debut. There are now more than 60 individual and team events.

OLYMPIC TORCH
As a symbol of international unity, the Olympic flame is taken by torch relay across national borders from the ancient site of Olympia in Greece to the Olympic venue. It is then used to ignite a flame that burns throughout the Games.

CHAMONIX CHAMPIONS
The Winter Games were first held at Chamonix in the French Alps in 1924, under the name of International Winter Sports Week. They were not officially recognized as a part of the Olympics until 1926.

GOLD STAR
Finnish ski-jumper Matti Nykänen won three gold medals at the 1988 games.

KEEP YOUR HEAD DOWN!
There are few sights in sports as spectacular as a bobsled on the run. Two-man or four-man teams push-start the sled for up to 50 m (165 ft), jump in as it gathers speed, and then, keeping their heads down below the sides of the sled to reduce wind resistance, hurtle down a treacherous icy track with high, curved sides. During a typical 60-second run the bobsled can hit speeds of 90 mph (145 km/h). The driver at the front steers the sled down the run, and the team with the lowest total time over four runs wins.

ICE DANCER
The most successful Olympic figure skater was Sonja Henie from Norway, who entered the 1924 Games aged 12 years. She won gold in three successive Games in 1928, 1932, and 1936, and went on to star in 11 Hollywood movies.

DOWNHILL DAREDEVILS

The hair-raising luge event was first included in the Winter Games in 1964. A luge is a lightweight sled, which is ridden down an icy track. Competitors race feet first and on their backs, steering with small body movements. Rubberized bodysuits and face shields help minimize air resistance.

SKI EVENTS

The traditional men's and women's Alpine skiing events are the downhill (shown here), slalom (in which skiers weave through a series of poles), giant slalom, and super-giant slalom, or "super-G"– a combination of downhill and slalom skiing. There is also freestyle skiing in which competitors race down bumpy slopes called mogul fields, or perform aerial tricks, spinning and somersaulting as they jump off snowy ramps.

The skier leans into the jump to streamline the body

JUMP FOR GOLD

Olympic ski jumpers launch themselves down a steep slide before taking to the air. Soaring up to 197 ft (60 m) off the ground, they can cover distances of more than 600 ft (185 m) before landing on the hill beneath, which is covered with well-groomed, packed snow. They are judged not only on the length of the jump but also on style and composure.

Downhill skiers use poles that curve around the body to help with turning and balance

BUILT FOR SPEED

Traditional, or long-track, speed skating is a graceful sport. Pairs of skaters race against the clock, using smooth, powerful strokes to propel them to speeds of up to 34 mph (56 km/h) over distances between 500 m and 10,000 m. Short-track skating, by contrast, is much more aggressive. Introduced as an Olympic sport in 1992, in Albertville, France, competitors race one another around a tight, 122 yard (111 m) track over distances of 500 m and 1,000 m for individuals, and 3,000 m for relay teams.

Steel skate blades

Falling rock

Snow and ice

Disaster!

THE WORST ACCIDENT in the history of climbing occurred on July 13, 1990, when a giant avalanche smothered 43 mountaineers at a camp 17,000 ft (5,200 m) up on Pik Lenin in Russia. Avalanches, landslides, and rockfalls are unpredictable and deadly. Although scientists know what conditions are likely to trigger these phenomena, and can put in place measures to control and stabilize them, the number of avalanche fatalities is still increasing. The major reason is a boom in mountain industries and recreation. Winter sports draw millions of people to the mountains every year, and to support these activities, more roads, buildings, and towns are forced into avalanche-prone areas.

Volcanic ash from Vesuvius set like cement around the body of this victim. The body decayed, leaving behind only its shape as a hollow in the rock

MOUNTAINS BLOW THEIR TOPS

Volcanic mountains are the sites of the most spectacular and devastating of natural disasters – volcanic eruptions. Perhaps the most famous was the one that shook Mt. Vesuvius near Naples in Italy in AD 79. When the long-dormant volcano erupted on August 24th, the residents of the Roman town of Pompeii were caught unaware. Hot ash and pumice rained down on the town until it was buried several feet deep. More than 2,000 people were overwhelmed and died.

ENGULFED IN MUD

The term "avalanche" is most often associated with snow, but it can also describe a flow of rock or mud. Mudflows like the one shown here in Caracas, Venezuela, usually strike without warning. Mud, rock, and soil form a porridge-like mixture that moves down a hillside at 6 mph (10 km/h) destroying all in its path. The "wave" of mud has been known to reach heights of 50 ft (15 m). Mudflows are most common in dry, hilly regions where prolonged rain has fallen.

Objects as large as boulders, cars, and even houses can be carried in the mudflow

AVALANCHE SURVIVAL
The best way to survive in an avalanche is to avoid it in the first place! If caught, experts advise "swimming" on the snow in a breast-stroke to keep on the surface. Once the avalanche stops, the snow solidifies, trapping unfortunate victims.

A large avalanche can unleash a volume of snow equivalent to 20 football fields filled 10 ft (3 m) deep

1 THE STARTING ZONE
It is hard to predict where an avalanche will occur. Temperature, wind direction, and snow conditions are all factors. Slopes of 30° to 45° are most at risk, especially after a heavy snowfall or sudden warm weather.

2 SLAB STORY
The most destructive type of avalanche occurs when a large, thick blanket of snow, compacted by wind or temperature, begins to slide over smooth ground or weak snow beneath. This is called a slab avalanche.

3 TRACKS FROM THE TOP
A huge mass of air is pushed in front of the falling snow and ice, creating a wind blast that can flatten trees. Large vertical swathes of forest missing from a slope indicate frequent avalanches in an area.

Saving the summits

SOIL ON THE MOVE
Cutting down forests on mountain slopes can lead to severe soil erosion. In Nepal, it is estimated that more than 300,000 tons of topsoil is washed off bare hillsides every year. It takes anything from 100 to 2,500 years for just 1 in (2.5 cm) of topsoil to be replaced by natural processes, so the effects of soil erosion are devastating both in the short term and long term.

THOUGH SEEMINGLY INDESTRUCTIBLE, mountains are easily damaged by human activities, such as mining, agriculture, and industry. One of the most serious problems is erosion. This occurs when forests are cut for timber, for firewood, or to make way for crops or cattle. Robbed of the protective layer of trees, thin mountain soils are washed away by the first heavy rains. The hillsides are left bare, while the silt clogs waterways in the lowlands, and empties into the sea where it chokes delicate coral reefs. Hunger for land, especially in developing countries, is a growing threat to mountain environments, but conservationists are helping to reverse the damage by replanting forests and encouraging thoughtful development.

MAKING LIGHT WORK
Solar panels harness the energy of the Sun to generate electricity or to heat water for washing and heating. They are well suited to the mountain environment because the sunlight at high altitude is very intense. Although expensive to buy, solar panels are cheap to run, and the energy they produce is clean and removes the need to fell more trees for firewood.

Fur turns pure white in winter

Arctic fox weighs about 9 lb (4 kg)

Arctic fox

TURNING UP THE HEAT
Mountain animals, such as the Arctic fox, are well adapted to cold conditions. But global temperature increases, caused by pollution from burning coal and oil, are making their habitats warmer. As this happens, they move to higher, cooler altitudes, and are replaced by species that thrive in warmer conditions. In the mountains of Scandinavia, the Arctic fox is slowly being replaced by the more common red fox.

SAVING THE FORESTS
In the Himalayas, large areas of forest have already been cut down, but there is hope for the future. There are a number of projects dedicated to replanting trees. The Sir Edmund Hillary Himalayan Trust nurtures almost 100,000 young trees each year, and has planted more than one million so far. Here saplings are being tended in a nursery at Ghandrung, a small village near Annapurna in Nepal.

Empty oxygen bottles await removal

GARBAGE DUMP
For 50 years climbers too exhausted to carry their trash down Everest have left it on the South Col. Most of it has blown away, but the heavy oxygen bottles remain. Today, expeditions follow strict environmental guidelines.

HOPE FOR THE PANDA?
Giant pandas are found in just six mountain forests in western China. Because their habitat is being damaged by development, there are only 700 to 1,000 of these animals left alive in the wild. Attempts to breed them in captivity have mostly failed, and the future of the species depends on the protection of their natural habitat.

Giant panda

Diet of bamboo shoots

Index

Acknowledgments

Dorling Kindersley would like to thank:
Bob Lawford and all at the Alpine Club, London, for help with archival material. Bob Sharp for loan of mountain rescue equipment and Chris Miller, at Lyon Equipment, for help with images of climbing equipment. John Cleare for assistance with picture research.
For editorial, design, and DTP help: Amanda Rayner, Sheila Collins, and Nomazwe Madonko.
Index: Lynn Bresler
For additional photography: Gary Ombler

Picture credits

a=above b=bottom c=center l=left r=right t=top

Advertising Archives: 2tr, 56cl. **Allsport:** 58–59; Jamie Squire 58cb; Mike Powell 59tl, 59cra; Olympic Museum 58tl, 58tr; Pascal Rondeau 56cra; Shaun Botterill 59br; Sylvie Chappaz 46bl. **Ancient Art & Architecture Collection:** R. Sheridan 33br. **Art Directors & TRIP:** H Rogers 35l; M. Jenkin 34cr; T. Bognar 20cl; W. Jacobs 26–27. **Ashmolean Museum:** 2tl, 3tl, 23c. **Birmingham Museum:** 2ct, 28cl. **Chris Bonington Picture Library:** 11t, 13r, 15tr, 46tr, 46–47, 50–51, 55tl, 55r; Doug Scott 15c; Dr Keuchi Yamada 3b, 10br; Hilary Boardman 11cr; Peter Boardman 15br; Leo Dickinson 47ca; Doug Scott 15c; Dr Keuchi Yamada 3b, 10br. **Bridgeman Art Library, London/New York:** Fitzwilliam Museum, University of Cambridge 40–41. **British Museum:** 37tr. **Bruce Coleman Ltd:** Mark Carwardine 18br. **Corbis UK Ltd:** 6c; W. Perry Conway 14tl; Hulton-Deutsch Collection 30tl; David Muench 8cr; Roger Ressmeyer 6b. **Dick Bass:** 10bl. **Mary Evans Picture Library:** 24cl, 29tl, 32br, 40bl, 41cr, 50tl, 52bl, 54l. **FLPA – Images of nature:** C. Carvalho 19tr. **Werner Forman Archive:** Courtesy Ariadne Gallery, New York 33tr; Nick Saunders/Barbara Heller 34tl. **John Frost Historical Newspapers:** 44tr, 45bl. **Gables:** 20br, 21r, 22tr, 23br, 23t. **Gettyone Stone:** David Hanson 24bl; Arnuif Husmo 25; Bob Torrez 2cr, 57c; Stuart Westmorland 16–17; Robert Yager 9tr. **Glasgow Museum:** 4tl. **Glenbow Museum, Calgary:** 30–31. **Robert Harding Picture Library:** Anchorage Museum 31br; Gavin Hellier 36–37. **Hornimann Museum:** 33l. **Hutchison Library:** H. R. Dorig 26tl, 29tc; Jeremy A. Horner 27tr; William Holtby 36tr. **INAH:** 28br, 32bl. **Lyon Equipment:** www. charlet-moser.com 4cr; www.marmot.com 49tl, 49bl; www.ortlieb.de 51tr. **John Cleare/ Mountain Camera:** 12bl, 13tl, 13tc, 42tr, 43b, 46–47b, 52–53,54b; Hedgehog House NZ 42–43; Colin Monteath 50c; Pat Morrow 11b, 63tr. **Museum of Mankind:** 26tr. **N.H.P.A.:** Laurie Campbell 16cl; Julie Meech 9cb. **S. Noel:** 42tl, 42clb, 43cl. **Oxford Scientific Films:** Sean Morris 16tl; Tom Ulrich 19cl. **Pitt Rivers Museum:** 27br. **Planet Earth Pictures:** 11cl; Nick Garbutt 62tl; Yva Momatiuk 10tr. **Popperfoto:** Werner Nosko 41br. **Rex Features:** M. Leon 60b; Today 53br. **Royal Geographical Society:** 43tr, 44l, 45tr, 45cl, 45br; Gregory 44cr, 45tl. **Royal Museum of Scotland:** 2c, 3tr, 4bl, 26bl, 28tr, 28bl, 29tr, 29cl, 29c, 29bl, 36tl. **Science & Society Picture Library:** National Railway Museum 38bl. **Science Photo Library:** David Parker 37tl. **Rebecca Stephens:** 10–11, 20cr, 37cr. **Still Pictures:** Roberta Parkin 61tr, 61cr, 61br, 61l; Hartmut Schwarzbach 62cr, 63tl. **Stock Shot:** J. Stock 36br. **Telegraph Colour Library:** Michael J. Howell 30bl; L. Lefkowitz 39b. **Jane Tetzlaff:** 22bl, 22l, 34bl, 39tl. **Topham Picturepoint:** 31tr, 31cr, 47br, 58c. **Williamson Collection:** 45bl.

Jacket credits **Chris Bonington Picture Library:** Dr Keuchi front cover cl. **John Frost Historical Newspapers:** back cover br. **Pascal Rondeau:** back cover cra. **Royal Geographical Society:** front cover bl, back cover br, tr. **Telegraph Colour Library:** front cover crb.

Every effort has been made to trace the copyright holders of photographs, and we apologize for any unavoidable omissions.

1
BIRD
2
ROCKS & MINERALS
3
SKELETON
4
ARMS & ARMOR
5
TREE
6
POND & RIVER
7
BUTTERFLY & MOTH
8
SPORTS
9
SHELL
10
EARLY HUMANS
11
MAMMAL
12
MUSIC
13
DINOSAUR
14
PLANT
15
SEASHORE
16
FLAG
17
INSECT
18
MONEY
19
FOSSIL
20
FISH
21
CAR
22
FLYING MACHINE
23
ANCIENT EGYPT
24
ANCIENT ROME
25
CRYSTAL & GEM
26
REPTILE
27
INVENTION
28
WEATHER
29
CAT
30
BIBLE LANDS
31
EXPLORER
32
DOG
33
HORSE
34
FILM
35
COSTUME
36
BOAT
37
ANCIENT GREECE
38
VOLCANO & EARTHQUAKE
39
TRAIN
40
SHARK
41
AMPHIBIAN
42
ELEPHANT
43
KNIGHT
44
MUMMY
45
COWBOY
46
WHALE
47
AZTEC, INCA & MAYA
48
BOOK
49
CASTLE
50
VIKING
51
DESERT
52
PREHISTORIC LIFE
53
PYRAMID
54
JUNGLE
55
ANCIENT CHINA
56
ARCHEOLOGY
57
ARCTIC & ANTARCTIC
58
BUILDING
59
PIRATE
60
NORTH AMERICAN INDIAN
61
AFRICA
62
OCEAN
63
BATTLE
64
GORILLA, MONKEY & APE
65
MEDIEVAL LIFE
66
FARM
67
SPY
68
RELIGION
69
EAGLE & BIRDS OF PREY
70
WITCHES & MAGIC-MAKERS
71
SPACE EXPLORATION
72
SHIPWRECK